Infinity: A Very Short Introduction

VERY SHORT INTRODUCTIONS are for anyone wanting a stimulating and accessible way into a new subject. They are written by experts, and have been translated into more than 45 different languages.

The series began in 1995, and now covers a wide variety of topics in every discipline. The VSI library now contains over 500 volumes—a Very Short Introduction to everything from Psychology and Philosophy of Science to American History and Relativity—and continues to grow in every subject area.

Very Short Introductions available now:

Available soon:

For more information visit our website

www.oup.com/vsi/

Ian Stewart

INFINITY

A Very Short Introduction

OXFORD
UNIVERSITY PRESS

OXFORD

UNIVERSITY PRESS

Great Clarendon Street, Oxford, OX2 6DP,
United Kingdom

Oxford University Press is a department of the University of Oxford.
It furthers the University's objective of excellence in research, scholarship,
and education by publishing worldwide. Oxford is a registered trade mark of
Oxford University Press in the UK and in certain other countries

© Joat enterprises 2017

The moral rights of the author have been asserted

First edition published in 2017

Published in the United States of America by Oxford University Press
198 Madison Avenue, New York, NY 10016, United States of America

British Library Cataloguing in Publication Data
Data available

Library of Congress Control Number: 2016955809

ISBN 978-0-19-875523-4

Printed and bound by
CPI Group (UK) Ltd, Croydon, CR0 4YY

I could be bounded in a nutshell
and count myself a king of infinite space.

William Shakespeare, *Hamlet* Act 2, Scene 2

Contents

List of illustrations

Infinity

Introduction

It may seem paradoxical to write a very short introduction to a very big concept, but infinity *is* paradoxical. It's also remarkably useful, and mathematicians and users of mathematics would be lost without it. However, it can also be dangerous, unless handled with considerable care. Philosophers and theologians have faced the same dilemma, though with different emphasis. It took more than two thousand years to learn how to handle the infinite without it exploding in our faces, and even then, it can still cause trouble.

The first recorded use of a specific word for the infinite is generally credited to Anaximander, a pre-Socratic Greek philosopher who flourished around 580 BC. His term *apeiron* can be translated in several ways—boundless, limitless, indefinite, infinite. His context was a search for the origin of all things, which he held to be an endless primordial mass. Being inexhaustible, *apeiron* could generate everything in existence without being used up. Exactly what he had in mind is unclear, but many scholars consider it to be a kind of primeval chaos that can be separated into the four ancient elements—earth, air, fire, water—from which, the Greeks believed, all else is formed.

Anaximander proposed that orderly reality had been created—*extracted* may be a better word—from formless chaos by pulling opposite qualities asunder. In this respect, *apeiron* resembles today's quantum-mechanical explanation of the origin of matter through the appearance of particle–antiparticle pairs, and is reminiscent of Galileo's paradox—an infinite set can be matched with a proper subset—and with the shenanigans that go on in Hilbert's hotel when infinitely many guests change rooms to accommodate a newcomer. Both can be interpreted as extracting something from an infinite set without anything being used up. Resolving this paradox was a key step towards a profound advance in our understanding of infinity: Georg Cantor's realization that some infinities are bigger than others.

The first known references to *mathematical* features of the infinite are the celebrated paradoxes of another pre-Socratic, Zeno of Elea, who lived between about 490 and 430 BC. The most famous is the fable of Achilles and the tortoise, in which the tortoise is given a head start. Achilles, though the faster runner, can never catch the tortoise, because by the time he reaches where it *was*, it has moved a little further on. So he has to perform infinitely many tasks before he can catch up, which allegedly is impossible. Zeno's paradoxes have a deceptive simplicity, but they challenge our intuition about space, time, motion, and causality.

Infinity lurks in the simplest, most mundane area of mathematics: arithmetic. When children first learn about numbers they often wonder what the biggest one is, usually settling for the biggest whose name they know—a hundred, or a thousand. But most of them quickly come to recognize that there is *no* biggest number, because adding one makes any number bigger. One way to say this is 'there is no largest number'. Aristotle called this kind of infinity 'potential infinity'. Another description, more contentious but richer in mathematical and philosophical promise, is: 'there are infinitely many whole numbers'. Aristotle called this kind of infinity 'actual infinity', but he didn't distinguish mathematics from reality the way we do now, so 'actual' is a misnomer.

Why do we need to think about the infinite—a concept we never encounter directly? There are many reasons. Even in elementary mathematics, we encounter aspects of infinity, for example when writing the fraction 1/3 as a decimal. To get an exact representation, the decimal must 'recur': repeat the same block of digits forever. More generally, our minds seem to *require* the idea that things might 'go on forever'—in space and in time, in the future and the past. Infinity is, perhaps, a mental default, a natural side effect of the pattern-seeking abilities of our minds. Evolution has moulded us to notice patterns in the external world, be they real or imaginary. Patterns have survival value. Going on forever without changing is perhaps the simplest pattern of all.

In consequence, we're happy to explain time as something that has always existed, and therefore has no origin. We find that more comfortable than time somehow *beginning*, even though that's what current cosmology proposes. We object to time starting by asking 'what came before?', failing to grasp that if time had a beginning, there was no 'before'. We prefer to think that space is infinite, and the universe extends without limit, because we imagine that if not, there must be a boundary—and we ask 'what lies beyond the boundary?' We're wrong on two counts. If the universe ends somewhere, there's nothing beyond, not even empty space. And the universe could be finite but unbounded.

Infinity—especially its temporal version, eternity—plays a significant role in much religious thinking. It's a standard topic in philosophy. It has intrigued artists as well as scientists. It sounds impressive, you can attribute all sorts of properties to it, and no one can prove you wrong unless your logic is in error. More positively, it's a fascinating concept, full of subtleties, logical pitfalls, puzzles, and paradoxes.

One of the greatest paradoxes of the infinite is that it's turned out to be extremely useful. As the inspiration behind calculus, it's taken humanity to the Moon, and flies millions of us across the globe every day. Mathematicians find it very difficult to get anywhere without infinity, even in areas of the subject such as combinatorics, which counts *finite* sets of objects. Patterns in these numbers can often be neatly packaged into a single infinite object called a generating function, which can then be manipulated to obtain useful information about perfectly finite things.

Mathematicians have even given infinity its own special symbol: ∞. There are also more recent symbols for specific *types* of infinity, such as $\aleph_0$ and ω, which we meet in Chapter 7. Perhaps the most important mathematical contribution to our understanding of the infinite is the realization that the same word 'infinity' can have many distinct interpretations. These can be defined rigorously, and their similarities and differences can be deduced logically from the definitions.

Although there exist philosophical views of what mathematics *should be* that forbid all reference to the infinite, virtually all practising mathematicians worldwide find the concept not just useful, but indispensable. However, there are also some intriguing scientific questions about physical infinity. For example: is the universe finite or infinite? What happens inside a black hole? Usually physicists interpret infinity as a sign that their theory has departed from reality, but many of them rather like the idea of an infinite universe. I'll examine the psychology behind this inconsistency in Chapter 6.

Infinity is a two-edged sword. Used with due caution, it opens up important methods such as calculus, upon which most of modern science is founded. Many of today's technological wonders were invented using some aspect of the infinite—even digital technology, which operates on finite binary numbers, but is built using materials science, optics, chemistry, and quantum physics—all involving the mathematics of the infinite in essential ways.

These triumphs notwithstanding, very minor changes to the way infinity is used can equally well lead to nonsense. And it's not always easy to distinguish a dividing line between the profound and the absurd. All of this makes infinity one of the most fascinating concepts ever invented. If 'invent' is the word.

Outline of the book

An introduction can open up some basic questions and answers, but it can only touch upon the deeper issues behind them. My main aim here is to get you thinking about those issues, and to raise awareness of the subtle distinctions that philosophers, theologians, and mathematicians have been forced to make when contemplating infinity. My viewpoint will be that of modern pure mathematics, which focuses on logical issues. Physics and applied mathematics often make less formal use of the infinite, but this isn't a comprehensive scholarly treatise, and I'll only skim the surface.

We therefore begin with a warm-up chapter, introducing nine typical examples of reasoning about the infinite—puzzles, paradoxes, even a few proofs. We discuss each of them briefly, and analyse whether the methods or the answers are logically acceptable. Some deserve further discussion, and we'll return to them in due course.

The second chapter raises some common misconceptions about infinity, and shows how infinity naturally appears in elementary arithmetic. The aim is to show how deeply embedded infinity is, even in basic areas of mathematics, and to clarify possible confusion about topics that we *think* we understand.

Chapter 3 focuses on historical attitudes to infinity, mainly in philosophy and religion, including Zeno's famous paradoxes. Infinity isn't a thing, but a concept, related to the default workings of the human mind. Zeno's paradoxes appear to be about physical reality, but they mainly address how we think about space, time, and motion. A central (but possibly dated) contribution was Aristotle's distinction between actual and potential infinity. Theologians, from Origen to Aquinas, sharpened the debate, and philosophers such as Immanuel Kant took up the challenge. Mathematicians made radical advances, often against resistance from philosophers.

Chapter 4 examines a logical counterpart of the infinite: infinitesimals. These are quantities that are infinitely small, instead of infinitely large. Historically, such quantities formed the basis of calculus, one of the most useful branches of mathematics ever invented. However, they caused considerable head-scratching, starting an argument that took about two centuries to resolve. This was achieved using a version of Aristotle's potential infinity—namely, potential *infinitesimality*, if there is such a word. (There is now.)

On several occasions I've rather casually called infinity a concept. It's not. It's a meta-concept: a jumble of more or less related ideas, masquerading under the same name. Much of the philosophical and mathematical fun comes from trying to tease the different meanings apart, and deciding which make sense, and why. A clear example occurs in Chapter 5, where the discussion takes a sharp left-hand turn into a different realm of the infinite: projective geometry. As Euclid insisted in one of his axioms, parallel lines never meet. But the painters of the Italian Renaissance, analysing perspective, stumbled across a rich vein of geometry in which it makes sense to insist that parallels do meet—at infinity. If you've ever stood at a railway station watching the tracks converge as they disappear into the distance, you've caught a glimpse of geometric infinity.

From mathematics we move to the real world, and Chapter 6 tackles questions such as 'is space infinite?' In many areas of physics, the presence of an infinite quantity (often called a singularity) is construed as a warning that the theory is losing touch with reality. For instance, according to classical ray optics, the intensity of light at the focus of a lens is infinite. The physical resolution of this difficulty involves replacing light rays by waves. In cosmology, however, the possibility of infinite space is more respectable.

Chapter 7 returns to the mathematics of infinity, discussing Cantor's remarkable theory of how to count infinite sets, and the discovery that there are different sizes of infinity. For example the set of all integers is infinite, and the set of all real numbers (infinite decimals) is infinite, but these infinities are fundamentally different, and there are more real numbers than integers. The 'numbers' here are called transfinite cardinals. For comparison we also mention another way to assign numbers to infinite sets, by placing them in order, leading to transfinite ordinals. We end by asking whether the old philosophical distinction between actual and potential infinity is still relevant to modern mathematics, and examining the meaning of mathematical existence.

Chapter 1
Puzzles, proofs, and paradoxes

To get us thinking critically and imaginatively about the infinite, here are some deductions and questions that use it. Some give the right answer, some don't, and some are plain baffling. Think about them before reading on. Compare them. Why do some make sense but others not?

Nine appeals to infinity

Largest number

Infinity (∞) is the largest number there is. So $\infty + 1 = \infty$. Subtract ∞ from both sides to get $1 = 0$.

Diagonal of a square

Imagine a regular 'staircase' along the diagonal of a unit square (Figure 1). The total length of this polygon—treads and risers—is 2, because the treads add to 1 and so do the risers. If the number of steps becomes infinite, and the steps become infinitely small, the staircase becomes the diagonal of the square. Therefore the length of the diagonal is 2.

Area of a circle

A circle is a curve formed by infinitely many infinitely short lines. Joining them to the centre as in Figure 2 creates infinitely many

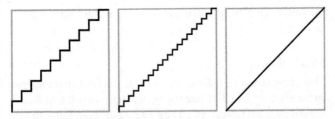

1. *Left and middle*: successive refinements of a staircase. *Right*: the limit with infinitely many steps.

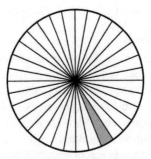

2. Infinitely many infinitely thin triangles: one shaded. (**Only 32 shown.**)

Puzzles, proofs, and paradoxes

infinitely thin triangles, each with perpendicular height equal to the radius r of the circle. Each triangle has area $\frac{1}{2}r.b$, where b is the length of the base, so summing them all, the area of the circle is $\frac{1}{2}r$ times its circumference. The circumference is $2\pi r$, so the area is $\frac{1}{2}r.2\pi r = \pi r^2$.

Light switch

At time 0, a light switch is off. After half a second I switch it on. A quarter of a second later I switch it off. An eighth of a second later I switch it on again. A sixteenth of a second later I switch it off, and so on. Each successive interval of time between moving the switch is half the previous one. After one second, is the light on or off?

Balls in the bag

I have infinitely many balls, numbered 1, 2, 3,..., and an empty bag. At time 0, I put balls 1–10 into the bag and take out ball 1. At time 1/2 second, I put balls 11–20 into the bag and take out ball 2. At time 3/4 second, I put balls 21–30 into the bag and take out ball 3. At time 7/8 second, I put balls 31–40 into the bag and take out ball 4, and so on. The number of balls in the bag increases by 9 each time. So after one second, how many balls are there in the bag?

One third in decimals

If we try to express 1/3 as a decimal, it can never terminate because 10 divided by 3 is 3 with remainder 1, so the calculation repeatedly gives 3s: 0·333 333... going on forever. If we stop at any specific place, say 0·333 333, the number is less than 1/3, because on multiplying it by 3 we get 0·999 999, which differs from 1 by 0·000 001. So is the infinite (recurring) decimal 0·333 333... smaller than 1/3, or exactly equal to it?

Squares and numbers

This extract from Galileo's 1638 *Discorsi e Dimostrazioni Matematiche Intorno a Due Nuove Scienze* (discourses and mathematical demonstrations relating to two new sciences) has been slightly edited for length.

Salviati: We cannot speak of infinite quantities as being the one greater or less than or equal to another. I take it for granted that you know which of the numbers are squares and which are not.

Simplicio: I am quite aware that a squared number is one which results from the multiplication of another number by itself; thus 4, 9, *etc.*, are squared numbers which come from multiplying 2, 3, etc., by themselves.

Salviati: Very well; and you also know that just as the products are called squares so the factors are called sides or roots. Therefore if I assert that all numbers, including both squares and non-squares, are more than the squares alone, I shall speak the truth, shall I not?

Simplicio: Most certainly.

Salviati: If I should ask further how many squares there are one might reply truly that there are as many as the corresponding number of roots, since every square has its own root and every root its own square, while no square has more than one root and no root more than one square...This being granted, we must say that there are as many squares as there are numbers because they are just as numerous as their roots, and all the numbers are roots.

Sagredo: What then must one conclude under these circumstances?

Salviati: So far as I see we can only infer that the totality of all numbers is infinite, and the attributes 'equal', 'greater', and 'less', are not applicable to infinite, but only to finite, quantities.

Hilbert's hotel

In a lecture in 1924 David Hilbert illustrated Cantor's theory of infinite numbers (transfinite cardinals) by imagining a hotel with infinitely many rooms, numbered 1, 2, 3,.... Suppose a new guest arrives when all rooms are already full. At first sight, the newcomer will have to find another hotel, but the manager has a brainwave. He asks every guest to vacate their room and move to the room whose number is one greater. That is, the guest in room 1 goes to room 2, the guest in room 2 goes to room 3, the guest in room 3 goes to room 4, and so on. They all move simultaneously. Now all existing guests still have a room, and room 1 is miraculously free for the new guest.

In a finite hotel this won't work: the guest in the room with the largest number has nowhere to go. But in Hilbert's hotel, there's no largest room number.

You might like to consider two further questions:

- Suppose the hotel is full and a coach arrives with infinitely many new guests, say in seat numbers 1, 2, 3,...on the coach. Can the hotel accommodate them all by moving guests around?
- What if infinitely many infinite coaches arrive? Again, assume the coaches are numbered 1, 2, 3,..., and so are the seats in each coach.

Grandi's proof of the Creation

In 1703 Guido Grandi published *Quadratura Circula et Hyperbolae per Infinitas Hyperbolas Geometrice Exhibitata* (quadrature of the circle and hyperbola exhibited by infinite geometric curves), in which he considers the infinite series

$$1 - x + x^2 - x^3 + x^4 - \ldots$$

By the binomial theorem this equals $1/(1+x)$. Set $x = 1$ to deduce that

$$1 - 1 + 1 - 1 + 1 - 1 + \ldots = \tfrac{1}{2}$$

On the other hand, we can group the terms as

$$(1 - 1) + (1 - 1) + (1 - 1) + \ldots = 0 + 0 + 0 + \ldots = 0$$

Therefore $0 = 1/2$, which Grandi interpreted as a proof that God can create the world from nothing. Another grouping is

$$1 + (-1 + 1) + (-1 + 1) + \ldots = 1 + 0 + 0 + \ldots = 1$$

so $0 = 1$, equally puzzling.

Solutions and comments

From the viewpoint of today's mathematics, most of these examples can be dealt with without introducing too many new ideas. Some need extra discussion, continued in subsequent chapters.

Largest number

The reasoning is clearly false, but why? One problem might be the assumption that infinity is a number, which in turn raises the question: what is a number? The deduction would be valid for a conventional number, so ∞ can't be a number in any conventional sense. However, mathematicians have defined less conventional meanings in which infinity is a (new kind of) number. There the statement '$\infty + 1 = \infty$' is mathematically acceptable, although a different symbol is normally used to make the context clear. What's *not* acceptable is subtracting ∞ from both sides of the equation, because subtraction can't be defined for infinite quantities if we want the usual rules of arithmetic to hold.

Diagonal of a square

To make sense of this example, mathematicians rephrase the argument in terms of a finite number n of steps, which *tends to* infinity. That's a fancy way to say 'remains finite but grows indefinitely large'. Whatever the value of n, the length of the staircase is 2. There's a well-defined limiting curve, and it is indeed the diagonal line. However, by Pythagoras's theorem the length of the diagonal is not 2, but $\sqrt{2}$. It's sometimes claimed that the limiting curve is not the diagonal, but an infinitely wiggly line that repeatedly crosses it. Not so. The length of the limiting curve is not the limit of the lengths of the staircases. That's all.

Area of a circle

The method described gives the right answer, and something very like it can be justified. Archimedes did so using a Greek method called exhaustion, but he assumed without proof that a circle has a well-defined area, as explained in Chapter 4. Today we usually resort to calculus instead. The idea is to use a finite number n of very thin slices, all exactly the same shape and size, straightened at their outer ends to form triangles. They don't cover the original circle exactly, but together they approximate its area very closely.

The area of each triangle is half its base times the perpendicular height, so the total area is half the perimeter times the perpendicular height. The perimeter is very close to the circumference of the circle, length $2\pi r$. The perpendicular height is very close to the radius r. So the total area is very close to $\frac{1}{2}.2\pi r.r = \pi r^2$. By estimating how close, and applying the logical foundations of calculus as in Chapter 4, it can be proved that the limit of the total area of the triangles, as the number n tends to infinity, is exactly πr^2. This limit is the *definition* of the area of the circle in calculus, avoiding the assumption that the area exists. In compensation, we must also prove that area, defined in this manner, has all the expected properties.

Light switch

Mathematically, the procedure defines the state of the switch for all times *less than* one second. This tells us nothing about the state after one second. Not all infinite processes have a sensible meaning, and this is one of them.

Physically, we would quickly end up trying to move the switch faster than light, which is impossible by relativity. Before that, friction would have melted the switch. Before *that*, the light bulb would probably have blown.

Balls in the bag

Infinitely many? Not so fast!

Please ignore the practical issues here. This is a hypothetical exercise in a non-relativistic world. The setting can be formalized to make mathematical sense.

At stage n the bag contains $9n$ balls, but we can't just let n tend to infinity to deduce that the 'final' number of balls is infinite. The limit of the number of balls in the bag is not the number of balls in the limit of the bag. In this respect, it's like the staircases and the diagonal of the square.

Actually, after one second there are *no* balls in the bag. To see why, observe that ball n is removed at the nth stage, and is never put back. Every ball goes into the bag, hangs around for a time, and then gets taken out again. When it all shakes down, the bag is empty.

One third in decimals

An infinite decimal can be given a rigorous logical meaning, see Chapter 4. That granted, suppose that

$$S = 0{\cdot}333\,333\,333\ldots$$

Then

$$10S = 3{\cdot}333\,333\,333\ldots = 3 + S$$

Therefore $9S = 3$, so $S = 3/9 = 1/3$.

Although stopping at any given stage yields a number less than $1/3$, the difference decreases rapidly the more decimal places you use. An infinite sequence that becomes arbitrarily small has limit *zero*. The apparent paradox arises because it never reaches that value.

Squares and numbers

It's remarkable that Galileo came so early to a conclusion that wasn't fully sorted out until Cantor developed his theory of counting for infinite sets over 200 years later: see Chapter 7. Adopting Cantor's view, most of what Salviati says at the end is correct, except that 'equal', 'greater', and 'less' *can* be applied to

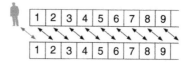

3. All move up one, and Room 1 is free.

infinite quantities. However, they don't behave exactly as they do for finite quantities, which you could argue is what Salviati really meant.

Hilbert's hotel

Hilbert's scenario amounts to a proof, within Cantor's framework for infinite sets, that if the 'number of elements' in the set of whole numbers (a transfinite cardinal) is denoted by $\aleph_0$, then $1 + \aleph_0 = \aleph_0$. (The symbol $\aleph$ is the Hebrew letter 'aleph'.) The underlying idea is to map the set **N** of whole numbers to its subset **M** of whole numbers greater than 1. This map must be a *one-to-one correspondence*, meaning that different elements of **N** map to different elements of **M**, and all elements of **M** arise from **N** in this way. Figure 3 shows how this can be done. The top line represents $1 + \aleph_0$, the bottom line represents $\aleph_0$, and the arrows prove they're equal.

Suppose an infinite coach arrives (Figure 4). The manager now moves each existing guest n to room $2n$, the even numbers. Then each coach passenger m can be assigned to room $2m-1$, the odd numbers. Again all guests can be accommodated. In Cantor's notation, this proves that $\aleph_0 + \aleph_0 = \aleph_0$.

To deal with infinitely many coaches, the manager assigns room numbers according to Figure 5, following the diagonal arrows and returning to the next room in the top row after reaching the end of each arrow. All guests can be accommodated, and this proves that $\aleph_0 \times \aleph_0 = \aleph_0$.

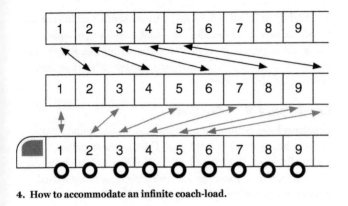

4. How to accommodate an infinite coach-load.

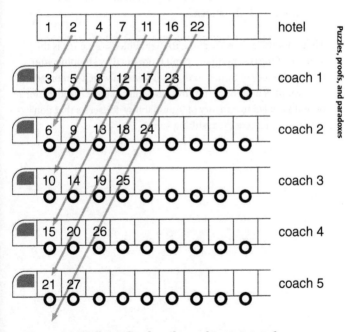

5. The manager's 'diagonal' order—the numbers 2–3, 4–5–6, 7–8–9–10, and so on slant to the left.

The slightly weird arithmetic for $\aleph_0$ makes sense if we interpret it naively as 'infinity'. If you add 1 to infinity, double it, or square it, you ought to get infinity again. Cantor's amazing discovery is that the arithmetic of transfinite cardinals is much richer than that. We'll see why in Chapter 7.

Grandi's proof of the Creation

Leonhard Euler performed a similar calculation around 1730 and was happy with 1/2 as the sum. Later mathematicians eventually decided that in order for an infinite series to have a meaning, it must *converge*: get arbitrarily close to some fixed number if you add enough terms together, see Chapter 4. This series converges when $-1 < x < 1$, but not when $x = 1$. So letting $x = 1$ is illegitimate.

That's not the end of the story, though. Euler's value 1/2 is the average of the two numbers 1 and 0 obtained by stopping after a finite number of terms, so there's a sense in which it represents the overall behaviour better than any other value. Considerations of that kind led to a theory of 'summability' for series that don't converge, and here the resulting sum is 1/2.

Chapter 2
Encounters with the infinite

We use the word 'infinity' rather casually in everyday speech, so it's worth clarifying a few basics before we plunge into subtler aspects of the infinite. I'll focus on two issues:

- Infinity is not just a synonym for a very big number. We often use it that way, for poetic or dramatic reasons, or just out of ignorance, but in mathematics and philosophy infinity is a different concept altogether: not a very large limit, but the absence of any limit.

- Infinity is not just some esoteric invention in advanced mathematics. We run into it quite early on at school level. The first important occurrence is not the absence of a largest whole number. We don't really need to know that, and teachers can easily slide round it if asked. The infinite raises its head in a more significant manner when we're taught about decimals and put them together with our previously learned concept of fractions.

Finite and infinite

I'm not going to start by *defining* infinity, because the word has many meanings, and I want to work my way towards these. As a working rule of thumb, a number (whole, fractional, decimal, whatever) is finite if it's smaller than some number in the familiar sequence 1, 2, 3,..., and infinite if not. (For negative numbers, make them positive first.) An object is finite if its size is finite, and infinite if not. So a circle is finite, but a line that goes on forever is not.

There are many measures of size, and the same object may be finite by one measure but infinite by another. A circle has finite circumference and area, but is composed of infinitely many points. The snowflake curve of fractal geometry is obtained from an equilateral triangle by repeatedly adding smaller equilateral triangles along the middle third of each edge (Figure 6, top). Its length is infinite but it encloses a finite area. The arc of the hyperbola $y = 1/x$ running from $x = 1$ to infinity (Figure 6, middle) has infinite length, and the area between it and the x-axis (shaded) is infinite. In 1644 Evangelista Torricelli proved that when this curve is spun about the x-axis to form a surface of revolution (Figure 6, bottom), it has infinite area but encloses a finite volume. In fact, the volume is exactly π. This surface is called Gabriel's horn or Torricelli's trumpet, and at the time it posed a serious challenge to mathematical intuition about the infinite.

Not just a big number

Standard dictionary numbers stop with centillion. This is 10^{303} in American and modern British usage, in which a billion is a thousand million. It's 10^{600} in the rest of Europe and by the older British convention in which a billion is a million million. Suggestions for extending the names lead to millinillion, which is 10^{3003}. This is considerably bigger than the famous googol, 10^{100}, but a lot smaller than a googolplex, which is $10^{\text{googol}} = 10^{10^{100}}$. Even the googolplex is tiny compared to infinity. Infinity is bigger than any specific whole number, whatever notational system we use and whatever new names we invent. In practice we run out of names before we run out of numbers.

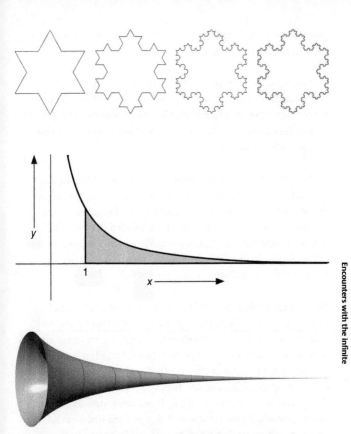

6. *Top*: successive stages in the construction of the snowflake curve. *Middle*: area under a hyperbola. *Bottom*: Gabriel's horn, also called Torricelli's trumpet, has infinite area but finite volume.

Archimedes understood this, and he wrote *Psammites* (sand reckoner) to disprove the assertion that the number of grains of sand on the surface of the Earth is infinite. It's certainly bigger than any number the ancient Greeks could name in their everyday language, but Archimedes considered that to be evidence for the paucity of common language. He wrote a pamphlet addressed to Gelo II, the King of Syracuse:

> I will try to show you by means of geometrical proofs ... that, of the numbers named by me and given in the work which I sent to Zeuxippus, some exceed not only the number of the mass of sand equal in magnitude to the Earth filled up, but also that of the mass equal in magnitude to the universe.

Archimedes then derived a finite upper bound on the number of grains of sand that the universe can contain, by combining two ingredients: a model of cosmology, and his novel method for naming very large numbers. He concluded that, in our terms, at most 10^{63} sand grains can fill the universe. With today's figure for the size of the observable universe, that becomes 10^{93}. Still finite.

Other cultures also took an interest in very large numbers. The Jain religion in India was founded around 600 BC, taking over from Vedic religions. Jains believe that all living creatures embody soul, which is immortal and perfect. Souls, including human ones, migrate to new creatures after death. To escape this endless cycle of transmigration, the devotee avoids any action that harms a living creature. Even swatting a fly is not permitted. Jainism has no concept of a divine creator or destroyer, and believes the universe to be eternal, both in its past and its future.

Jain cosmology involves a very long period of 2^{588} years, roughly 10^{177}. But the Jains acknowledged a clear distinction between 'very large' and 'infinite'. The *Anuyoga Dwara Sutra*, probably dating from about 100 BC, discusses a hemispherical trough with diameter 100,000 *yojanna*, about 1 million kilometres. 'Fill it with white mustard seeds counting one after the other. Similarly fill up with mustard seeds other troughs ... Still the highest enumerable number has not been attained.'

Mathematicians have devised notational systems for numbers far larger than anything that Archimedes or the Jains contemplated. But these, too, are finite.

Infinity is bigger.

Infinity in arithmetic

Mathematical infinity first appeared in connection with the *whole numbers* 1, 2, 3, 4,... and so on, often also called the counting numbers. Mathematicians generally throw 0 into the mix as well, placing it at the front, which gives the *natural numbers*. Historically, the next extension of the number system led to fractions like 1/2 and 4/15, followed by the introduction of negative numbers −1, −2, −3,... and −1/2, −4/15. The positive and negative natural numbers are the *integers*, and when positive and negative fractions are included as well we get the *rational numbers* or plain *rationals*.

All these system are infinite, in the sense that there is no largest counting number: however large n may be, $n+1$ is larger. The other two systems contain even more numbers, so they must also be infinite. However, we don't run into these instances of infinity in any essential manner when we learn arithmetic. The first meaningful encounter with infinity, for most of us, is when we start learning about the decimal system. Even there, we start with finite decimals, such as 3·14 or 1·41421 (which happen to be approximate values for π and $\sqrt{2}$). Arithmetic with finite decimals is essentially the same as arithmetic with integers; we just have to learn where to put the decimal point.

Pandora's box opens up when we put fractions and decimals together, and ask what 1/3 looks like as a decimal. To a good approximation, the answer is 0·333. To a better approximation, 0·3333. To a better approximation still, 0·33333. If you want a *really* close approximation, keep appending more 3s. However, none of these approximations is exact. To see why, multiply them by 3, obtaining

$$0.999 \qquad 0.9999 \qquad 0.99999$$

An exact value would give the answer 1, but these are smaller. They differ from 1 by

$$0.001 \qquad 0.0001 \qquad 0.00001$$

and although these numbers rapidly become very small, none of them is zero.

Although five or six decimal places are generally enough for practical calculations, exact representations are desirable for mathematical purposes. Otherwise the decimal system would miss out a lot of interesting and useful numbers. Fortunately, there's a way round the difficulty with 1/3: the concept of a *recurring* decimal. This repeats the same pattern of digits indefinitely, perhaps after a finite number of exceptions that don't fit the pattern. One notation is to put dots over the digits at the front and end of the repeating sequence. So

$$0 \cdot \dot{3} = 0 \cdot 33333 \ldots$$

going on forever, and

$$0 \cdot 6 \dot{5} 7142 \dot{8} = 0 \cdot 6 \ 571428 \ 571428 \ 571428 \ldots$$

It then turns out that every fraction can be represented *exactly* by a recurring decimal, or a finite one if the decimal representations stops, as it does for 1/2 = 0·5. (You can consider this to be followed by a recurring block of zeros if you prefer.) Conversely, every recurring (or finite) decimal represents a fraction. The two displayed numbers are 1/3 and 23/35, for instance.

Irrational numbers

A recurring decimal may 'go on forever', but we can prescribe it in its entirety in a finite manner—as I just have. The spanner in the works was thrown by an ancient Greek; legend has it that he was Hippasus of Metapontum. He was a member of the Pythagorean cult, which believed that everything in the universe rests on numbers. At that time, the word implied whole numbers, or fractions formed from whole numbers. One of the triumphs of Pythagoreanism was what we now call Pythagoras's theorem: the square on the hypotenuse of a right-angled triangle is the sum of the squares on the other two sides. Hippasus had been musing about the diagonal of a unit square. By Pythagoras's theorem, the square of its length must be $1^2 + 1^2 = 2$, so the length of the diagonal is the square root of 2. Hippasus proved that the square of a rational number cannot equal 2. You can get close, indeed as close as you wish, but you can't represent the diagonal exactly using rational numbers.

This was a blow to Pythagoreanism's deepest belief. According to the legend (for which there's no evidence) Hippasus unwisely announced his result on board a boat crossing the Mediterranean, and his fellow travellers were so incensed they threw him overboard. The result, however, is undeniable: in modern terminology, the square root of 2 is *irrational*. The Greeks got round this difficulty, for theoretical and philosophical purposes, by abandoning numbers and working geometrically with lengths, areas, volumes, and angles. When decimals were invented, however, the awkward status of $\sqrt{2}$ once more came to the fore.

It follows from Euclidean geometry, with a bit of extra assistance to fill in some logical gaps in Euclid's presentation, that every geometric length can be expressed as an infinite decimal. (I'll show you how later.) The decimal expansion of $\sqrt{2}$ starts like this:

$$1·414\ 213\ 562\ 373\ 095\ 048\ 801\ 689$$

There's no obvious pattern that repeats the same block of digits forever. Appearances might be misleading, because in principle that block might be very large, but a recurring pattern represents a rational number, whereas $\sqrt{2}$ is irrational. So the digits of $\sqrt{2}$ go on forever, but we can't specify them by a straightforward rule that lets us predict exactly what, say, the millionth digit must be. For recurring decimals, we can: it's 3 for 1/3 and it's 1 for 23/35.

However, there is a rule. Each successive digit is the largest one that makes the square of the result less than 2. We can use this rule to calculate as many digits as we wish. In practice there are more efficient methods, but this one works. The standard arithmetical method for finding square roots is really just an efficient variant. In 1768, Johann Lambert proved the long-standing conjecture that our old friend

$$\pi = 3{\cdot}141\ 592\ 653\ 589\ 793\ 238\ 462\ 643$$

is irrational.

As soon as we reach the stage in our mathematical education at which geometry and decimal notation collide, we face the notion of a decimal expansion that continues *forever*, but need not recur. These 'infinite decimals' are (a conceptual notation for) the *real numbers*. Unless the number can be written as a fraction with denominator a power of 10, stopping at some finite stage is always an approximation.

Numbers through the microscope

I claimed that Euclidean geometry lets us prove that any length has an infinite decimal expansion. Let me show you what I have in mind, using $\sqrt{2}$ as an example since that can be constructed geometrically by ruler and compasses. Figure 7 illustrates the first four steps. At the top, we locate $\sqrt{2}$ on the number line between 1 and 2, corresponding to the first digit, namely 1. In the second

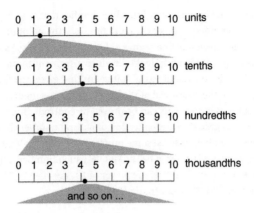

7. Constructing the digits of $\sqrt{2}$.

we magnify the interval from 1 to 2 tenfold, and locate $\sqrt{2}$ on this subdivision of the line. It lies somewhere between 4 and 5, corresponding to the next decimal approximation 1·4. In the third row we magnify the interval from 4 to 5 tenfold, and locate $\sqrt{2}$ on this further subdivision. It lies somewhere between 1 and 2, corresponding to 1·41. Yet another subdivision locates it at 1·414, and if the process were to continue we would obtain successive digits 2, 1, 3, and so on.

You won't find decimals in Euclid, but Book VI, Proposition 2 of the *Elements* is a geometric construction to subdivide a line segment into any number of equal pieces. So this process is consistent with Euclidean geometry.

Given a line of length π, the same process would give its successive digits. Although π is not constructible with ruler and compasses, the Greeks knew other ways to construct such a line. Indeed, any line segment leads to an infinite decimal expansion. Conversely, we can reverse-engineer the whole approach to find out exactly where on the number line any given infinite decimal appears.

27

I've left out one important point, which implies that the decimal expansion obtained in this manner determines the point uniquely. Namely, that given any line segment, however short, some such subdivision of a line of unit length leads to a segment that's shorter. This is equivalent to the statement that given any finite number (here the reciprocal of the length of that segment) some power of ten is larger. This can be proved by contradiction provided we agree that any collection of natural numbers has a smallest member. Assume some number exists that is larger than any power of 10. Let n be the smallest such number. Then $n-1$ is less than or equal to some power of 10, say $n-1 \leq 10^k$. Therefore $n \leq 10^{k+1}$, a contradiction.

The condition that any collection of natural numbers has a smallest member is called the well-ordering principle. Informally it's obvious: pick any number in the collection. If that number is the smallest, we're done. If not, we have only a finite number of candidates to check: the natural numbers less than our chosen one. In a formal treatment of the logical foundations of mathematics, we make the well-ordering principle an axiom.

My proof makes one hidden assumption: the reciprocal of any non-zero positive number is finite. We have to take that as an axiom as well. In a number system that includes infinitesimals, it's false: see Chapter 4.

Discrete and continuous

One of the great dichotomies in mathematics is the distinction between discrete objects and continuous ones. The natural numbers are discrete: each is separated from all the others by a definite gap. There's no natural number strictly between 1 and 2, or between 1066 and 1067, for that matter.

The real numbers are continuous. Given any positive real number, however small, we can find a smaller one: just halve it. Any interval of real numbers that contains more than one point can be subdivided into smaller intervals. Between any two distinct real numbers there lies at least one more real number; indeed, infinitely many.

Rational numbers have an uneasy existence between these two extremes. They're not discrete: you can find a different rational as close as you wish to any given rational. Between any two distinct rational numbers there lies at least one more rational number; indeed, infinitely many. Despite that, the rational 'number line' has gaps. It fails to contain $\sqrt{2}$ and π, for example. So it's not really continuous, either.

If we try to rework Euclid's geometry using only rational numbers, we run into the problem that Hippasus pointed out. In fact, we run into subtle consequences that are counter to our geometric intuition. Figure 8 shows circle centre O, with radius OA = 2. The bisector of this segment is B, and the perpendicular bisector is the line BC, meeting the circle at C.

How long is BC? Pythagoras's theorem implies that $OB^2 + BC^2 = OC^2$, which equals OA^2, which is 4. That is, $1 + OB^2 = 4$, so $OB^2 = 3$. However, there is no rational number with this property: $\sqrt{3}$ can be proved to be irrational. So

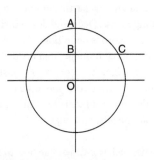

8. **Perpendicular bisector BC of OA meets the circle at C. Or does it?**

in the world of rational geometry, in which the only meaningful lengths are rational, the point C *does not exist*. The perpendicular bisector of a radius passes through the circle without actually meeting it. It 'squeezes' through a point-sized gap in the circumference of the circle.

This feature of the rationals reveals a subtlety in the traditional distinction between discrete and continuous number systems. There, continuity is usually conflated with infinite divisibility: any line segment, however short, can be subdivided into smaller pieces, as in the analysis of infinite decimals in Chapter 2. However, infinite divisibility does not imply continuity. The rationals are infinitely divisible, but they have gaps.

Euclid's *Elements* tacitly assumes this kind of thing can't happen. When David Hilbert studied the foundations of geometry in 1899, he found a large number of unstated assumptions of this kind. Euclid did a good job for his historical period, but by more modern standards his axiomatic treatment has many flaws.

In the 19th century, similar issues appeared in the traditional approach to real numbers and indeed integers. In his 1893 *Was Sind und Was Sollen die Zahlen?* (what are numbers and what should they be?), Richard Dedekind pointed out that everyone was assuming the truth of statements such as $\sqrt{2} \times \sqrt{3} = \sqrt{6}$, but no one had proved them. This particular one can be polished off using Euclid's geometry, but Dedekind developed a more general approach, based on a rigorous definition of a real number as a *section*: two sets of rational numbers that divide the line into a left-hand region and a right-hand region. For instance, the left-hand region of the section corresponding to $\sqrt{2}$ consists of all positive rationals whose square is less than 2; the right-hand region consists of all the other rationals.

Once sections are defined, you can define how to add them or multiply them, and prove that the standard laws of arithmetic hold. In this manner, Dedekind showed that once you have rational numbers, you can *construct* real numbers from them. At a price, however. A section is a pair of *sets* of rationals, and these sets are infinite. When you do arithmetic with such sets, you're conceptually working with infinite objects. Today's mathematicians have got used to this way of thinking, and are undisturbed by its philosophical overtones. Philosophers, on the whole, found it worrisome, and most of them argued against it. The argument eventually stopped because both sides lost interest. In Chapter 3, we'll examine some of the difficulties that they grappled with.

Chapter 3
Historical views of infinity

The history of humanity's concepts of the infinite, and the uses we've made of them, goes back over 2500 years to Anaximander's *apeiron*. Three main intellectual areas were involved: theology, philosophy, and mathematics (with recent interventions from theoretical physics, which I'll subsume under the 'mathematical' heading). A comprehensive treatment would be a major undertaking, so I'll summarize a few key figures and ideas. On the theological side, I'll restrict the discussion to Christianity, and merely try to give some flavour of the theological issues that were disputed and debated.

The development of mathematics since Anaximander's time can be divided into four main periods and locales. Initially, the main action was in Greece, as exemplified by Euclid's *Elements* and the philosophical works of Aristotle. For the next millennium it shifted to China, India, and Arabia. From 1400, following the Renaissance, most major advances in mathematics were made in Europe. By the 20th century the mathematical community had gone global. This is a very broad-brush description of a complex train of events, and throughout there were contributions from other regions and cultures.

Prior to the 20th century, religion and philosophy had a substantial influence on mainstream mathematical thinking.

The three areas were closely intertwined, often to an extent we now find surprising. Christian theology revered Aristotle and made his understanding of the infinite a keystone in its thinking about the nature, and existence, of God. The infinitude of the natural numbers became core evidence in theological debates. Philosophers discussed foundational issues in logic and mathematics, and mathematicians took inspiration from their conclusions. Individual mathematicians struggled to reconcile their mathematical discoveries with their personal beliefs.

At the start of the 20th century, with the development of an axiomatic basis for the foundations of mathematics, these links began to fall apart. Axiomatic set theory became too technical to appeal to philosophers, who mostly abandoned the philosophy of mathematics. The main exceptions were Bertrand Russell and Ludwig Wittgenstein, who disagreed with each other. Ironically Russell, ably assisted by Alfred North Whitehead, was one of the main people responsible for the deeply technical nature of the set-theoretic foundations of mathematics. Most mathematicians ceased to pay much attention to what philosophers said, especially when (as with Wittgenstein) they told the mathematicians they were doing everything wrong. Religion lost much of its political clout; throughout most of the developed world religious belief declined, though to differing extents in different countries. In particular, mathematicians no longer felt constrained by the teachings of the Church.

The adoption of precise axiomatic foundations for mathematics clarified the logical issues considerably, without necessarily resolving them all. Infinity was still puzzling, but at least we knew what we were talking about and why it was puzzling. Alongside these developments came a new viewpoint on mathematical existence. There's no need for mathematical concepts to be direct models of reality, or indeed to be related to reality at all. Mathematicians should consider themselves free to introduce new concepts, provided they don't create logical contradictions and can be related to existing concepts. As Cantor remarked, 'The essence of mathematics is its freedom.'

Warning: may contain infinity

The examples in Chapter 1 illustrate the power and dangers of the infinite as a framework for the advancement of mathematics. Arguments that seem almost identical can be valid in one context but fallacious in another. Historically, subtle distinctions of this kind often emerged from mathematical and philosophical controversies. In mathematics the concept of 'infinity' is neither predetermined nor unique: instead, it depends on the context and is defined according to the logical requirements of that context. Philosophers, too, have distinguished different interpretations of the infinite.

The central mathematical issue is whether familiar properties of finite objects and processes remain valid for infinite ones. Infinity is not alone in this respect; negative numbers are a case in point. The first historical record of negative numbers is the Chinese *Jiŭzhāng Suànshù* (nine chapters on the mathematical art), dating from the Han Dynasty (202 BC–220 AD) but probably going back much earlier. By 400, Chinese and Indian mathematicians made free use of negative numbers, but tacitly assumed they obey the same basic arithmetical laws as positive numbers. When complex numbers, in which minus one has a square root, were introduced, the same tacit assumption was made, but it was all terribly mysterious. Eventually mathematicians learned to define these extensions of the number system abstractly, to list explicitly the basic rules required to work with them, and to prove that the extended number systems did or did not satisfy any particular rule. Dedekind's use of sections of the rational numbers to define the real numbers is an example.

Logical treatments of infinity followed the same pattern. Initially, there was a naive assumption that basic features of finite processes would automatically be valid for infinite ones. An example is Gottfried Leibniz's 'law of continuity', which, in a letter to Pierre Varignon in 1702, he summarized as: 'the rules of the finite are found to succeed in the infinite'. Then there was a period of confusion when it turned out that assumptions of this kind were sometimes wrong. Finally, clarity emerged when the concepts were defined logically, the required properties were stated explicitly, and these properties were proved or disproved.

Preconceptions of the human mind

I suspect that our cognitive processes predispose us towards concepts of the infinite, because we naturally extrapolate simple patterns. This ability offers many evolutionary advantages—forecasting the seasons and the weather, avoiding predators by observing their hunting methods, understanding how plants grow in order to cultivate them. But it also leads us to extrapolate patterns from a small amount of evidence, and infinity is such an extrapolation.

Our mental images of space and time are affected by this kind of extrapolation, and to some extent reflect how our brains process images and events. Euclid's geometry is founded on lines and points, which are among the basic structures that the visual cortex extracts by processing incoming sensory data. Because there's no particular limit to the length of a line, the simplest assumption is that there's *no* limit. For two thousand years it was assumed that Euclidean geometry is a true representation of nature; indeed that it's the only possible geometry. In fact, it's neither; but it does embody an idealization of the simple patterns that our visual system presents to us.

Similarly, our sense of the passage of time leads us to arrange events in linear order. We perceive no beginning; memories of the past just fade out the further back we go; we also become aware that events happened before we were born. Our personal sense of time passing also seems to have no end, because when it does end, we cease to exist; moreover, we understand that most things will continue long after we're dead. Since we find it difficult to comprehend time stopping, we assume that it never will. The belief that 'it must continue' underpins all religions that believe in an afterlife.

Space going on forever is *infinite*; time going on forever is *eternal*. The former carries connotations of 'very large'. The latter of course refers to an endless period of time, but it has a second connotation: stability. Anything that lasts an eternity is considered, in some sense, to be unchanging. As a result, our default image for infinite time is subtly different from that for infinite space. This distinction is reinforced because objects extend across space, but processes *happen* in time. An object is a finished thing, existing and complete. A process ceases to be a process when it stops; all that is left is the result of the process. But a process can also be ongoing, able to continue; and, as already explained, if we see no good reason for it to stop, then we naturally imagine it goes on forever, and think it's eternal.

The mathematical formalism of natural numbers, integers, real numbers, and so on, explicitly builds this assumption into the definitions. However, it's open to challenge, as Zeno realized.

Zeno's paradoxes

The first known mathematical treatment of infinity occurred in the work of Zeno of Elea around 450 BC, and it has come down to us second-hand through Simplicius of Cilicia and Aristotle's *Physics*. Zeno discussed four paradoxical arguments about motion. Two explicitly rest on the infinite, and the third can be viewed as being about infinitesimals. The fourth is more obscure.

Achilles and the tortoise

Achilles has a race with a tortoise. He runs ten times as fast as the tortoise, so for fairness he gives it a head start—say 100 metres in modern units. By the time he reaches where the tortoise started from, the tortoise has moved a further 10 metres ahead. By the time he reaches that point, the tortoise has moved a further 1 metre ahead. By the time he reaches that point, the tortoise has moved a further 1/10 metre ahead...and so on. In order to catch up with the tortoise, Achilles has to pass through infinitely many points, but it's impossible to complete an infinite number of tasks.

Or, as Aristotle says in *Physics*: 'In a race, the quickest runner can never overtake the slowest, since the pursuer must first reach the point whence the pursued started, so that the slower must always hold a lead.'

The dichotomy

In order to traverse a given distance, we must first traverse half of it. But before we can do that we must traverse half of that first half, and before doing so, half of *that*...We have to complete an infinite number of tasks even to get started.

Or, as Aristotle says in *Physics*: 'That which is in locomotion must arrive at the half-way stage before it arrives at the goal.'

The arrow

In order for an arrow to move, it must change its position. But at any instant, the arrow must be stationary, since no time passes. If it's stationary at every instant, it can't move.

Or, as Aristotle says in *Physics*: 'If everything when it occupies an equal space is at rest, and if that which is in locomotion is always occupying such a space at any moment, the flying arrow is therefore motionless.'

The stadium

Aristotle's *Physics* again:

> Concerning the two rows of bodies, each row being composed of an
> equal number of bodies of equal size, passing each other on a
> race-course as they proceed with equal velocity in opposite
> directions, the one row originally occupying the space between the
> goal and the middle point of the course and the other that between
> the middle point and the starting-post. This...involves the
> conclusion that half a given time is equal to double that time.

This one is far more obscure than the other three, and not
obviously about the infinite. More plausibly, it's about motion
when both space and time are discrete. Aristotle dismissed it as
an evident fallacy. Kevin Davey's article in Further Reading gives
a detailed analysis, so I'll say no more about it.

Discussion of Zeno's paradoxes

Achilles and the tortoise

Simplicius tells us that Diogenes the Cynic refuted Zeno's
arguments by standing up and walking. This early example
of experimental disproof shows that something must be
wrong—either with the argument, or with how the assumptions
about motion relate to reality—but it doesn't tell us what.

Zeno's description of the paradox, and his argument, require
both space and time to be infinitely divisible. That is, no specific
smallest quantity exists, and any non-zero quantity can be made
smaller while remaining non-zero. In short, both space and time
are considered to be continuous. (In much of the philosophy,
infinite divisibility and continuity are conflated, although the
rational and real number systems show they're different, see
Chapter 2. I won't disentangle the distinction here because it
doesn't affect the discussion.) Aristotle pointed out that as the

distance to be covered decreases, so does the time taken to traverse it. In this continuum model of motion, a moving body *does* perform infinitely many tasks in a finite time.

Suppose for definiteness that Achilles runs at 10 metres per second, while the tortoise moves at 1 metre per second (fast for a tortoise; I'm being generous). We can do a mathematical Diogenes: after 20 seconds, Achilles has gone 200 metres while the tortoise has gone 20, so Achilles is 80 metres in front. He must have overtaken the tortoise along the way. We can find out exactly when. Solving the equation

$$10t = 100 + t$$

gives $t = 11\frac{1}{9}$. At that instant, both Achilles and the tortoise have reached the same point, $111\frac{1}{9}$ of a metre from where Achilles started.

The same answer arises if we follow Zeno's reasoning step by step. Achilles moves a distance

$$100 + 10 + 1 + 1/10 + 1/100 + \ldots$$

while the tortoise moves

$$10 + 1 + 1/10 + 1/100 + \ldots$$

Both take the same time:

$$10 + 1 + 1/10 + 1/100 + \ldots$$

From a modern viewpoint, these are convergent series, whose respective sums are $111\frac{1}{9}$, $11\frac{1}{9}$, $11\frac{1}{9}$.

There's a philosophical issue here, not completely answered by this calculation. During that time, both contestants have moved infinitely many times, thereby performing infinitely many tasks. Causality becomes awkward in such circumstances. However, in this case we can argue that although each task takes a non-zero time, the times decrease so rapidly that *all* of the tasks are completed in a finite total time.

We can avoid such considerations by raising a logical objection. Zeno's argument proves that Achilles does not catch the tortoise at an infinite number of specific times. However, that doesn't rule out the possibility that he catches it at some other time—and a continuum model of space and time shows that he does. This resolution of the paradox relies on making sense of a continuum, itself a nontrivial task. Dedekind's construction of the real numbers is one way to go.

If we use the continuum of real numbers to model both space and time, any interval of space contains infinitely many locations, and any interval of time contains infinitely many events. Motion combines both. *Any* movement of an object across a finite distance in finite time requires passing through infinitely many intermediate locations and infinitely many intermediate events. That's how the model behaves.

The dichotomy

The logical structure is similar to the first paradox, but now the instants of time under consideration converge to the start rather than the end. If it takes 10 seconds for the arrow to reach its mark, we're asked to consider times $10, 5, 5/2, 5/4, \ldots$. These form a decreasing infinite sequence converging to zero. Provided time is continuous, we can calculate the time at which the arrow reaches each point.

Again, motion involves carrying out infinitely many tasks. These tasks are not ordered like the positive integers, occurring in turn as the first, second, third, but again the argument considers only specific intermediate points ('tasks'). A continuum contains infinitely many others, in a more complex order than that of the positive integers or the negative integers. The paradox doesn't correctly represent the relation between space and time in a continuum model of motion.

The arrow

Many resolutions of the Arrow paradox have been proposed. At its heart is the problem of representing time as both a continuously flowing variable and as a succession of instants of zero duration. Aristotle wrote: 'Time is not composed of indivisible nows any more than any other magnitude is composed of indivisibles.'

This statement is at variance with the modern view of the continuum of real numbers, where time *is* composed of infinitely many indivisible nows (points), and the same goes for space. What matters is how these points combine to make a continuum, and this is not achieved by ordering them like a discrete system, such as the positive or negative integers. 'Succession' is the wrong image.

Thinking about the physics of motion, and how it relates to the usual continuum models, focuses attention on the assumption that an instantaneous snapshot of a moving object is indistinguishable from that of a stationary one. There are several objections to this. One, proposed by Peter Lynds, is that instants of time and instantaneous values of variables don't physically exist. However, they do in the mathematical models used by physicists, and the paradox is about those models. Bertrand Russell suggested that motion depends on observing an object at two times or positions, not just one. At any fixed instant it appears to be stationary, but if it's somewhere else at a later instant, it must have moved. Furthermore, between those instants, it must occupy every intermediate point in space (assuming for simplicity that it moves along a line).

Nick Huggett's resolution is to question Zeno's assumption that an object that is instantaneously in the same position as it would be at rest, must actually *be* at rest. One way to make sense of that is to recast the argument in the context of Hamiltonian systems, a general theory of mechanical systems that Sir William Rowan Hamilton developed from an earlier idea of Joseph-Louis Lagrange. In this formulation, the state of a point particle or a body, at any given instant, is determined not just by its position, but also by its momentum. Momentum is mass times velocity. A stationary body at a given position has zero momentum. A moving body at the same position has non-zero momentum. An instantaneous snapshot shows their positions to be the same, but can't distinguish their momenta. To do that, we must compare two snapshots at distinct instants, and see whether the particle has moved. So what Zeno is missing is the possibility of a 'hidden variable' that differs from location, and distinguishes a moving particle from a stationary one. This resolution can be seen as a formalization of Russell's and Huggett's views.

Philosophers and the infinite

For a long time after Zeno, the mathematics of the infinite faded into the background. Its philosophical and religious connotations did not. A key figure in the philosophy of infinity is Aristotle, who tackled the topic in *Physics* and *Metaphysics*. In the first, he relates infinity to nature, whose main features are change and rest. Change is continuous, hence infinitely divisible, so nature leads inevitably to contemplating the infinite. Studying the infinite is therefore justified as a necessary precursor to the study of nature. The infinite must exist, in some sense, for otherwise there would be 'many impossible consequences', such as time having a beginning and an end, or some lines being indivisible, contrary to Euclid.

The key question is then: in what sense does infinity exist?

It might exist in actuality: something infinite existing as a completed object. Alternatively, it might exist potentially: as a process that can always be extended, but which at any stage remains finite. In books 4 and 5 of *Physics*, Aristotle demolishes, to his own satisfaction, the possibility of actual infinity. Book 6 polishes off the argument: infinity can exist only potentially.

The simplest example of potential infinity is the process of counting. No matter how large a number you've reached, there's always a next one. The process has no limit, but at no stage does it reach an infinite number.

Euclid was clearly aware of this distinction. In his *Elements*, the theorem that we now state as 'there are infinitely many prime numbers' appears in a different form. Book IX, Proposition 20 states: 'Prime numbers are more than any assigned multitude of prime numbers.' That is, the process of listing prime numbers can be continued indefinitely. This is a statement about potential infinity. Nowhere does Euclid contemplate the 'object' comprising *all* prime numbers. He just proves that however many you have, you can construct another one.

However, there's a problem with Aristotle's conception of potentiality, as he himself admitted. He repeatedly makes statements like 'one thing after another is always coming into existence', with the emphasis on future continuation. That's all very well for counting, primes, extending geometric line segments, and so on. But it ran into trouble when confronted with Aristotle's belief that the universe has existed for an eternity, so time has no beginning. If so, surely the events that have happened in the past—all the past instants of time, the number of times the celestial spheres have rotated—constitute an *actual* infinity. You can't get them by allowing time to continue; you would have to run it backwards. So they don't fit the definition of potential infinity.

We don't know who raised this objection; it could even have been Aristotle, as devil's advocate. Later, around 550, John Philoponus argued against the Neoplatonist Proclus in his *De Aeternitate Contra Proclum* (on the eternity of the world against Proclus). Proclus followed Aristotle in asserting that the world had no beginning. Philoponus argued that this would make the past history of the world actually infinite. Simplicius then pointed out that Aristotle had already demolished that claim by sliding neatly past this objection. Past events can't constitute anything actual, because 'the parts that are taken do not persist'. They've come and gone. 'Actual' would require them to exist *here and now*.

Even so, this objection, says John Bowin in *Aristotelian Infinity*, 'caught Aristotle completely off guard, since his theory of the potential infinite was clearly devised to explain [infinite succession]'. Aristotle's answer tacitly invokes a third kind of infinity, neither actual nor potential. His deduction that actual infinities are impossible assumes infinity is either actual or potential, and these are mutually exclusive. So his claim to have proved that all infinities are potential has a logical gap.

Philosophers traded blows over the infinite for centuries, mainly going over the same ground in new ways. William of Ockham is famous for 'Ockham's razor': entities should not be multiplied beyond necessity. He wrote: 'Every continuum is actually existent. Therefore any of its parts is really existent in nature. But the parts of the continuum are infinite because there are not so many that there are not more, and therefore the infinite parts are actually existent.' Here he seems to be making a very fine distinction about the meaning of 'infinite', rather than about existence. Something can be actual, and infinite, without being actually infinite.

In his 1690 *An Essay Concerning Human Understanding* John Locke, a leading empiricist philosopher, stated that all human ideas stem from sensory perceptions. Since our senses are finite, so must our perceptions be. Since we can't perceive infinity, it doesn't exist:

The infinity of numbers, to the end of whose addition every one perceives there is no approach, easily appears to any one that reflects on it. But...there is nothing yet more evident than the absurdity of the actual idea of an infinite number. Whatsoever positive ideas we have in our minds of any space, duration, or number, let them be ever so great, they are still finite; but when we suppose an inexhaustible remainder, from which we remove all bounds, and wherein we allow the mind an endless progression of thought, without ever completing the idea, there we have our idea of infinity.

This is Aristotle's potential/actual distinction again. Locke made the further observation that our thinking about infinity is contradictory: 'Let a man frame in his mind an idea of any space or number, as great as he will, it is plain the mind rests and terminates in that idea; which is contrary to the idea of infinity...a supposed endless progression.' This is a category error. Our idea of something need not be the same as the thing itself. Our idea of a cat is not a cat, but a representation of a cat. We can *represent* infinity by something finite.

Kant's view was almost the exact opposite. The 1781 *Critique of Pure Reason* lists four 'antinomies': pairs of mutually contradictory beliefs, which in his view arise whenever the human mind attempts to grasp reality. The first antinomy bears directly on the infinite, contrasting two opposing views. Either the world had a beginning in time and is limited in space, or it's eternal and infinite. Reality, said Kant, transcends the mind, which is limited by our senses. Therefore the mind cannot grasp the true nature of reality. Kant's view is easier to understand in the context of space:

Space is not an empirical concept which has been derived from outer experiences. For in order that certain sensations be referred to something outside me...the representation of space must already underlie them. Therefore, the representation of space cannot be obtained through experience from the relations of outer appearance; this outer experience is itself possible at all only through that representation.

In Kant's jargon, our minds have synthetic a priori knowledge of the properties of space. Among those properties are infinite extent and infinite divisibility.

The danger with this view is that it promotes conceptual notions of space, such as Euclid's, above observation. Kant considered Euclidean geometry to be necessary and universal. We now know that it's neither, not even within mathematics. Empirical observation has shown that real space isn't Euclidean.

Infinity in Christian belief

Some version of the infinite occurs in many religions, but I'll focus solely on Christianity to keep the topic within bounds. As Philoponus illustrates, the philosophy and mathematics of infinity became intimately entwined with early Christian beliefs. In medieval times, the notion that God has no limits became entrenched; it was pretty much the definition of the Deity. Man is ephemeral, mortal, with limited powers and knowledge; God is eternal, immortal, omnipotent, and omniscient.

The Bible provides less support for these beliefs than we might expect. In the King James version, 'infinity' never occurs, and 'infinite' appears just three times. The most relevant is Psalms 147:5: 'Great is our Lord, and of great power: his understanding is infinite.' However, Job 22:5 reads 'Is not thy wickedness great? and thine iniquities infinite?' which suggests the metaphorical meaning 'very large'. 'Eternal' appears more often, but most references are legalistic: eternal covenant, eternal agreement. Only a few concern attributes of the Deity. Other words with similar meanings, such as 'everlasting', 'immortal', also occur, but these too are rare.

Early theologians seem not to have considered God to be literally infinite. Around 200 AD, in *De Principiis* (on first principles), Origen, the first Christian theologian of repute, maintained that God's power is finite. The reason is that perfection can't have blurred edges. Its limits must be sharp. Latin *perfectus* means 'complete'. If God's power were infinite, it would be incomplete, hence imperfect.

The infinitude of God becomes explicit around 395 AD, when Eunomius argued that Christ, as the son of God, is subordinate to God. The son was created, therefore had not always existed, therefore was not divine. He further argued that Creation as a whole is finite, so the son is finite—again not divine. The Council of Constantinople formally condemned this Eunomian heresy in 381. Gregory of Nyssa provided a lengthy counterargument, which I won't attempt to summarize, to demonstrate that Eunomius's claims fail if God is infinite. By the time of Augustine of Hippo, around 400 AD, the infinitude of God had become fundamental to Christian theology. He even gave a mathematical proof: 'Let us then not doubt that every number is known to him "of whose understanding," the Psalm [147:5] goes, "there is no set number".'

Emphasis on God being infinite was reinforced by medieval attempts to prove His existence. In the *Proslogion* of 1077–8 Anselm of Canterbury presented what's now called the ontological proof of God's existence. (Ontology is the philosophy of pure being.) Anselm himself had a more personal objective: his book describes how, by meditating, he became convinced that God exists. His argument can be summarized, in broad terms, as follows. Consider the most perfect possible being. Since a being that exists is more perfect than one that does not, the most perfect possible being must exist.

This sketch may not do full justice to the subtlety of the thinking, but it captures the main line of argument. The view that God has no limitations is a direct consequence of the ontological argument. The most perfect possible being cannot have any specific limitation, for the same reason there's no largest whole number. If you state a limit, something greater is conceivable, which would describe a more perfect being.

It's difficult not to feel that Anselm's argument gets something from nothing. Merely contemplating a hypothetical being leads, without any empirical evidence, to a real one. Kant attacked it as fallacious in his *Critique of Pure Reason* of 1781, arguing that existence is not a logical predicate—a property that something can possess or lack. If it were, the statement 'God exists' becomes 'There is a God, and He has the property of existence.' Sounds fine, but by the same reasoning 'God does not exist' becomes 'There is a God, and He had the property of non-existence', which is self-contradictory.

Mathematically, the fallacy is clear. You can't infer properties of an object from its definition until you've proved such an object exists. For example, consider the definition 'the largest positive integer'. Here's a proof, by contradiction, that this integer is 1. Let x be the largest positive integer, and suppose that $x > 1$. Then $x^2 > x$, contrary to the definition of x. Therefore $x = 1$. The fallacy is that no such x exists in the first place. What we've proved is actually 'if x exists then $x = 1$'. In logic, there are two ways for 'if P then Q' to be true. One is that Q is true. The other is that P is false. So the best we can infer is that either $x = 1$ or x does not exist. Anselm's ontological argument similarly allows us to conclude that *either* the most perfect being exists, *or* the most perfect being does not exist. It would be difficult not to agree, but it gets us precisely nowhere.

Such objections notwithstanding, many religious devotees accepted the ontological argument, and its corollary: no limits. This led them to believe not just in a very powerful deity, but an omnipotent one; not a very knowledgeable deity, but an omniscient one; not a very long-lived deity, but an eternal one; not a very extensive deity, but an infinite one.

Around 1260–70 Thomas Aquinas offered a different proof of the existence of God, also depending on infinity. His *Summa Theologiae* and *Summa Contra Gentiles* discuss five such proofs, of which the second relates to causality. He asserted that an infinite chain of causality is impossible, so there must be a First Cause. This is God.

Suppose we take any particular event—say, getting out of bed this morning. Among its prior causes is the manufacture of the bed. This traces back to the felling of the tree that provided the wood, then to a seed from the previous generation of trees, and so on. Aquinas argued that this kind of reverse sequence of causes can't explain anything if it just goes back forever. The 'explanation' would have no basis: the whole sequence should also have a cause. This echoes a statement I made earlier, and it's closely related to Zeno's dichotomy paradox.

Philosophers have argued for and against the 'First Cause' proof for a variety of reasons. One objection is that everything in existence is supposed to have a cause... *except* the First Cause. Why the special pleading? Another problem is causality itself: what is it, and why do we think all events have causes? A third is the assertion that an infinite chain of causality is impossible, which is assumed as an axiom without justification. A fourth is the identification of the First Cause with the God of Christianity.

A more mathematical one, apparently unnoticed, is that even if a First Cause exists, it need not be unique. Causality corresponds roughly to a 'partial order', in which a given entity can be larger than another, smaller, or the two may be incomparable. For a partial order, minimal elements need not exist, but even when they do, they need not be unique. Without uniqueness, there's no rationale for identifying just one of a multitude of First Causes with any particular entity, real or hypothetical.

Minimal elements always exist for a partial *well*-ordering, in which any descending sequence must stop after finitely many steps (so there's no chain of causality reaching infinitely far into the past—Aquinas's axiom), but minimal elements still need not be unique. Uniqueness does hold for a total order: given any two elements, one of them is greater than the other. But causality isn't a total order.

Modern era

Today's mathematicians think about the infinite in a rather different way, and seldom draw Aristotle's distinction between actual and potential. Mathematics is conceptual, both in its objects and its processes. Psychologically, these are distinct; mathematically, they're two sides to the same coin. Today's notion of mathematical existence is not the same as that of physical existence.

Aristotle put mathematical infinities such as numbers into the same category as 'all the men that have ever lived'. To the ancients, many mathematical concepts were 'real'—in an idealized Platonic sense. Plato's theory of forms asserts that the highest kind of reality consists of abstract forms, or ideas, and that the material world is an imperfect image of the ideal one. Euclid's geometry was thought to be the true geometry of space, albeit using ideal perfect forms such as points with location but not size, and circles that were perfectly round, drawn with lines of zero thickness. Reality, inked on papyrus or scratched in sand, was a pale shadow of the ideal. Properties of real objects could be deduced by considering their ideal versions, which were simpler. (Imagine trying to define which grains of sand constitute the point of intersection of two lines drawn as slightly wobbly grooves; especially since the lines and points are places where the sand is no longer present.) Everything Aristotle and many of his successors said about the infinite was mixed up with this confused view of mathematical reality.

Today's mathematicians don't consider the distinction between actual and potential infinity to be important, because mathematical objects are 'actual' only on a conceptual level. Infinity isn't the problem, though it does add to the confusion. What about 'two'? I can show you two cats or two chairs—but I can't show you the *number* 'two'. Holding up a piece of paper with '2' on it doesn't work; that's a symbol for the number, a numeral. Not the actual (!) number. That said, mathematicians would answer the 'past time' objection to potential infinity by allowing processes to run backwards as well as forwards. A mathematical process is a sequence of steps, each 'following' the other in logical, not temporal, succession. In the progression of years '2016, 2015, 2013, 2012, ... ' each year succeeds the other in the sequence, but *precedes* it in historical time.

Even if you insist on maintaining temporal order, similar reasoning applies. The number of events since any specific past time can be made bigger by starting earlier. If things come one after the other, then the other comes before the one. In practice this is pretty much the position that Aristotle adopted.

Could there be a largest number?

Lofty philosophical arguments and mathematical abstractions sometimes lose contact with empirical reality. In both, we fondly imagine that however big a number someone writes down, we can always write down a bigger one. However, as a practical matter, this isn't true.

The googol, 10^{100}, is 1 followed by a hundred 0s. It takes only a couple of minutes to write this out in full in base-10 notation. It's easy to write down the result of adding one, or to put an extra zero at the end to multiply it by ten. We really can write down a bigger number. But compare the googol to its big brother, the googolplex $10^{10^{100}}$, which is 1 followed by a *googol* of 0s. The human lifespan is too short to write this out in full in base-10 notation, or even to make a significant dent in the task. Moreover, the entire global supply of paper and ink, from now until ten billion years hence when the Sun expands into a red giant, would be inadequate to record that number. Of course we can write $10^{10^{100}} + 1$, but for any pre-specified notational system, there comes a point when it's not possible to write a bigger number down. There's not enough time, or not enough room. Our finite world, though gigantic, *can't* continue the process as we naively imagine.

Our use of infinite decimals—indeed, very long finite decimals— also falls foul of reality. Our mental image of space is an infinitely divisible continuum. Any interval, however short, can be subdivided into ten shorter ones, as in Chapter 2. But matter starts to be become indivisible when we get down to the scale of an atom, and in quantum mechanics, space is indivisible on the scale of the Planck length, which is 16×10^{-35} metres. On theoretical grounds, no measurement smaller than about one tenth that size is possible. So, in the context of actual physical measurement, numbers with more than 35 or 36 digits after the decimal point have no sensible meaning.

These remarks have an interesting implication. All of the numbers that anyone has ever used, be it for mathematics, science, medicine, or buying food, using any notation yet invented, are smaller than some specific number. I have no idea what it is, and writing it down would immediately destroy that property, but it must exist. So in practice, only numbers smaller than that bound have ever been needed. Absolutely no activities that depend on numbers, in the whole of human history, would change if we had limited ourselves to this finite range of numbers.

So why do mathematicians insist that the range of numbers must be infinite? One reason is the unconscious assumption that if a simple pattern persists for a long time, it must persist forever. A second is that when mathematicians started to formalize the processes of counting and arithmetic, they realized that everything is simpler if we assume from the start that certain arithmetical rules are *universal*. One of these is that $n+1$ is always bigger than n. If we abandon the convention that there are infinitely many whole numbers, the traditional rules of arithmetic, hence also algebra, don't work.

It's not *impossible* to set up a finite version of arithmetic with a very big largest number, but it's inelegant and difficult to work with. Mathematicians prefer their patterns to be universal in scope, so they embrace the infinitude of the whole numbers. Infinity is *simpler* than some specific but inexplicit very large number.

Chapter 4
The flipside of infinity

We now turn from the infinitely large to the infinitely small. Three examples in Chapter 1 (diagonal of a square, area of a circle, one third in decimals) are in this category. Each describes a process in which a geometric object or a number is repeatedly subdivided, or approximated ever more accurately by finer and finer structures, and the result is then made exact by considering an infinitely fine—*infinitesimal*—subdivision.

The ancient Greeks were excellent logicians, and recognized that this method is fallacious when expressed in such terms. However, they found a rigorous way to make sense of it, which they called *exhaustion*. Eudoxus used this method to put the theory of proportion on a sound logical basis when the lengths involved are incommensurable—in effect, to deal with irrational numbers, although the Greeks preferred to reason in terms of lengths of lines, not numerical measures of those lengths.

We'll take a brief look at exhaustion, and then progress, by way of calculus, to the modern concept of a limit, which abolished infinitesimals. Then we'll see how they were reinstated.

Proof of Archimedes's theorem

Archimedes didn't use π explicitly. Instead, he proved that the area of any circle is equal to its radius multiplied by half the circumference. If we define π as the ratio of the circumference to the diameter, this result is equivalent to the usual formula πr^2. It can be motivated by cutting the circle into ever-thinner slices, and thinking about a limiting case of 'infinitely many infinitesimal slices' as in Chapter 1. But this approach lacks logical rigour. Instead, Archimedes used exhaustion, based on sequences of approximating polygons whose areas and perimeters were known. One sequence approximates the circle from inside, the other from outside.

Let A denote the radius multiplied by half the circumference. Then the following statements are mutually distinct and together *exhaust* all possibilities:

(1) The area of the circle is greater than A.
(2) The area of the circle is less than A.
(3) The area of the circle is equal to A.

Instead of trying to prove (3) directly, the method *disproves* both (1) and (2), using proof by contradiction. Logically, only (3) remains.

The 'outside' sequence of approximating polygons is defined by circumscribing a regular hexagon round the circle, and repeatedly bisecting angles to create circumscribing regular polygons with 12, 24, 48, 96...sides. Figure 9(a) shows the first two stages in this process—later stages are too close to the circle to draw clearly. The geometrical details of the disproof of (1) are complicated, but the basic idea is simple. If (1) holds, the area of the circle exceeds A by a specific amount $d > 0$. Each external polygon has greater area than the circle, so its area is also greater than $A + d$. However, if the number of sides is sufficiently large, the area of the polygon

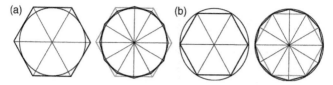

9. The first two stages in approximating a circle. (a) from the outside. (b) from the inside.

can be proved to be less than $A + d$. This contradiction disproves (1). A similar argument using the 'inside' sequence of polygons disproves (2), Figure 9(b). Now statement (3) follows.

The main practical deficiency of exhaustion is that you have to know the correct answer A to set up the trichotomy. The main theoretical deficiency is that you have to know that the quantity you're seeking exists. The Greeks assumed that every shape—in particular, every circle—has a well-defined area and a well-defined perimeter (or circumference). Much later it turned out that this assumption involves subtle features of analysis, but that with enough care and effort areas and lengths of many shapes—though not all—can be defined and various sensible properties proved. But by the time this was sorted out, better methods than exhaustion had emerged.

Calculus and its precursors

We no longer use exhaustion to study areas and volumes, because a simpler and more general technique was developed: calculus. The history of calculus is complicated, with a series of precursors and a full-blown controversy over who deserves the credit for bringing the subject to fruition: Leibniz or Isaac Newton. (The consensus is 'both'.) Initially, the logical formulation was rather vague, making intuitive use of infinitesimals without providing clear definitions. This was typical of mathematics until the 1800s in any case—not just for infinitesimals, but also for numbers, functions, and other less esoteric concepts.

Leibniz and Newton unified two distinct branches of the subject:

- Integral calculus, which calculates lengths, areas, volumes, and similar quantities.
- Differential calculus, which calculates the instantaneous rate of change of some quantity; for example, acceleration is the rate of change of velocity. Geometrically, the aim is to find the tangent to a curve at a given point.

Integral calculus (say for an area) proceeds by dividing an approximation to the area into pieces with simple shapes, calculating the area of each piece, adding the results, and then making the pieces arbitrarily small and their number arbitrarily large to remove any error, as in Figure 10 (left). Differential calculus divides the change in the quantity concerned, over a small interval of time, by the length of that interval; then the interval is made arbitrarily small, as in Figure 10 (right). So both processes involve 'infinitesimal' quantities, and integral calculus also involves infinite ones (the number of pieces). I'll discuss these processes in more detail shortly, but we need the gist of them now to see how the historical ideas relate to the final outcome.

Precursors to calculus abound. Democritus, a Greek philosopher who flourished around 400 BC, is mainly remembered for the theory that everything is made from indivisible atoms. But he was also among the first to discover that the volume of a cone is one third the area of its base times its height. He probably obtained this result by slicing the cone into infinitesimally thick circular sections parallel to the base, treating each as a very thin cylinder, and adding their volumes. Unlike most of his successors, Democritus had reservations about the logic of this procedure, but it gave the right answer.

In the medieval period, the main contributions came from China, India, and the Middle East. Around 500 AD Zu Gengzhi stated that if two bodies have the same cross-sections when

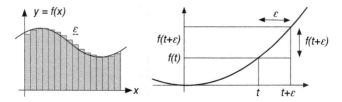

10. *Left*: area under the graph of *f* approximated by rectangles of width ε.
Right: rate of change of function *f* over a small interval of length ε.

sliced by equidistant parallel lines (planes), then their areas
(volumes) are equal, which is a variant of Democritus's method.
Around 1000 AD Alhazen (Abū al-Haytham) made another
prediscovery of concepts we now associate with integral calculus.
He used formulas for sums of squares and sums of fourth powers
of integers to find the volume of a paraboloid—in effect,
calculating integrals.

In the 14th century Madhava of Sangamagrama, the leading figure
in the Kerala school of mathematics, developed the technique of
expressing a function as a power series, discovering what later
became known as the Taylor series of a function, credited to
Brooke Taylor. He applied his method to state power series
expansions for trigonometric functions. Power series later became
an important application of calculus, and the basis for analysis,
especially with regard to complex numbers.

In 1635 Buonaventura Cavalieri published *Geometria
Indivisibilibus Continuorum Nova Quadam Ratione Promota*
(geometry developed by a new method through the indivisibles
of the continua). Like Democritus and Zu, he treated areas
and volumes as sums of infinitely many infinitely thin parallel
slices, which he called indivisibles. He used this idea to find the
area under the curve $y = x^n$, an early example of integration.
His method is called 'Cavalieri's principle', and he made effective
use of it.

A typical example of the principle shows that the area of a triangle is half the base times the perpendicular height. First, observe that this is true for a right-angled triangle, because two copies fit together to form a rectangle (Figure 11, left). Cavalieri's principle extends the result to an arbitrary triangle. Slice it into infinitely many horizontal lines and slide them sideways so that their left-hand edges form a line at right angles to the base (Figure 11, right). This converts the triangle into a right-angled one with the same base, perpendicular height, and area. Euclid gave a different, less contentious, proof of this result.

Cavalieri's principle also gives correct formulas for the areas and volumes of polygons, circles, cylinders, cones, spheres, and more esoteric objects. However, it has to be used with care. For example, as well as sliding the lines to the left to create a right-angled triangle, we could also move them downwards by halving their height above the base. Every line in the original triangle still matches a unique line of the same length in the right-angled triangle, but the area is halved.

Pierre de Fermat, famous for his conjecture in number theory, took Cavalieri's ideas further, defining a concept that he called *adequality*: differing only by an infinitesimal error. In the 17th century John Wallis, Isaac Barrow, and James Gregory put the two branches of calculus together. By 1670 Barrow and Gregory had proved (though not rigorously) that integration is the reverse of differentiation, a result often called the fundamental theorem of calculus.

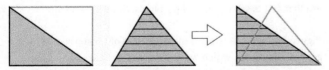

11. Example of Cavalieri's principle. *Left*: area of right-angled triangle is half that of the rectangle. *Right*: slide the slices to change the shape to a right-angled triangle with the same area.

The rigorous justification of Cavalieri's principle, carefully stated, is most simply obtained through calculus. Leibniz and Newton put all of these ideas together into a systematic package. They used different notation but their main results and concepts were very similar; not because either stole the other's ideas, as was later claimed, but because the subject naturally fits together in only one way. Today, elementary differential calculus mostly uses Leibniz's notation $\mathrm{d}y/\mathrm{d}x$ for the derivative (rate of change) of quantity y with respect to quantity x, and $\mathrm{d}^2y/\mathrm{d}x^2$ for the second derivative. But traces of Newton's notation still remain, such as $\dot{y}$ and $\ddot{y}$ for the same things, or f' for the derivative of a function f. The term 'calculus' is Leibniz's; Newton called the subject 'fluxions'.

Leibniz took a 'pure mathematical' viewpoint and was mainly interested in the philosophical implications of calculus. Newton's approach gave birth to theoretical physics and applied mathematics. His formulation relied on physical intuition, and he used calculus to answer a wide range of basic questions in the physical sciences. Ironically, the extensive applications of calculus to physics over the next century were mainly discovered in continental Europe, not Britain, and used Leibniz's notation.

Limits

It's possible to become proficient in calculus by learning a lot of rules and practising them, but understanding why those rules are correct is another matter. We can learn by rote the rule that the derivative of the function $f(x) = x^2$ is $f'(x) = 2x$. We can even use that fact to solve practical problems. But why is it true?

Figure 10 (right) illustrates the definition of the derivative for a general function f, which we now take to be $f(x) = x^2$. The calculation of f' considers a small increment from x to $x+\varepsilon$, so that the function changes from x^2 to $(x + \varepsilon)^2$. Then the difference in the x-values is ε, while the difference in the function is $(x+\varepsilon)^2 - x^2$. Their ratio ('average rate of change' over the interval ε) is

$$\frac{(x+\varepsilon)^2 - x^2}{\varepsilon} = \frac{2\varepsilon x + \varepsilon^2}{\varepsilon} = 2x + \varepsilon$$

which is *not* exactly $2x$.

Now comes the sleight of hand. If ε becomes very small, the expression $2x + \varepsilon$ gets very close to $2x$. The easiest way to see this is to set $\varepsilon = 0$, in which case only $2x$ remains. However, Bishop Berkeley pointed out (with some heat) that the previous step then involved the fraction $0/0$, which is meaningless.

A similar trick is used in integral calculus, with a similar objection. To find the area under the graph of a function, approximate it by a series of rectangles of width ε and then let ε become very small (Figure 10, left). Berkeley would object that if ε is not zero the area is wrong, but when $\varepsilon = 0$ each rectangle has area zero, so the total is zero too, which is also wrong.

Leibniz attempted to deal with these issues by considering ε to be infinitesimal, a concept that he explained in detail. For the derivative, it could then be neglected to leave $2x$. Something similar dealt with the integral. Newton used a physical image instead: ε is not a fixed quantity, but one that *flows towards* zero without ever reaching it. Then $2x + \varepsilon$ flows towards $2x$.

It probably didn't help that Newton used the symbol o where I've used ε. This allowed Berkeley, in effect, to accuse him of confusing o with 0. In his 1734 book *The Analyst*, Berkeley scathingly referred to o as the 'ghost of a departed quantity', claiming that calculus obtained correct results through compensating errors. In a way, he was right, but he was too busy indulging in theological point-scoring to ask himself a more important question: *why* do the errors always compensate? If you can explain that, with a strong enough guarantee, they're not errors at all.

Mathematicians generally ignored Berkeley, not because he was wrong but because they found the entire argument irrelevant. The results emerging from calculus included powerful insights into heat, sound, elasticity, gravity, electricity, magnetism, and fluid flow. Even if there *were* tiny errors in the sums, they could be made a lot smaller than measurement errors in experiments.

When a rigorous formulation of calculus was devised in the 19th century, the subject became known as analysis. Augustin-Louis Cauchy very nearly sorted out the logical foundations when he was developing a theory of analysis using complex numbers and functions in place of real ones. In his view, an infinitesimal is a *variable* quantity that approaches (but need not reach) zero; in effect, a sequence of numbers a_n that becomes arbitrarily small if n is large enough. An example is $a_n = 1/n$, which is never zero, but can be made as small as we please. Then, writing ε for this sequence, the expression $2x + \varepsilon$ becomes the sequence $2x + a_n$. This differs from $2x$ by a_n, which is infinitesimal in Cauchy's sense, so 'in the limit' we get $2x$. What we *don't* do is just set $\varepsilon = 0$.

The concept of a variable is itself informal, however, so this approach falls short of modern levels of rigour. Eventually Bernhard Bolzano and Karl Weierstrass devised the formulation we use today. A quantity $g(x)$ *tends to a limit L* as x tends to a fixed number a if $f(x)$ can be made arbitrarily close to L by making x sufficiently close to a. To formalize this statement we specify *how* close. Let ε be any positive number. Then there must always exist some positive δ, depending on ε, such that whenever $|x-a|<\delta$ we have $|f(x)-L|<\varepsilon$. This '$\varepsilon - \delta$' definition of a limit is precise, makes no explicit use of physical imagery such as 'flowing', and makes no mention of infinitesimals. Everything that appears is an ordinary real number. We don't even insist that ε or δ is small.

That's just where the main implications come into play. If δ works for some ε, it also works for anything larger.

Aristotle would recognize a cunning application of potential infinity (more accurately, potential infinitesimality) here. We don't make ε infinitesimal. We take it to be any positive real number. We think of ε as being small, but the main point is that whatever size it is, we can always make it smaller. Then we must make δ smaller too, but that's permitted. It has to be; if we specified δ once and for all at the start, we could make ε so small that the condition on L failed.

With definitions such as these, Bolzano and Weierstrass turned calculus into analysis in its modern sense. The infinitesimal was banished, even as informal motivation. In its place was a complicated form of words, liberally sprinkled with 'for all' and 'there exists' quantifiers, affectionately called 'epsilontics' by irreverent mathematics students. With an effort, you could learn to master the language, and analysis fitted together logically and it all made sense. As generation upon generation of students went through the process of getting used to it, the bad old days of the infinitesimal faded from mathematical memory.

Infinite series

With the rigorous formulation of the concept 'limit', calculus powered ahead to become just one part of a much broader area of mathematics, *analysis*. Limits resolved several other basic issues to do with infinity and infinitesimals, by recasting them in finite terms. Aristotle would have been proud, because the essence of this resolution is a move from actual to potential infinity. The $\varepsilon - \delta$ definition of a limit is based on a process that assigns to *any* specific finite positive ε a specific finite positive δ.

Limits also make sense of infinite series. The definition is modified slightly; in place of a real number δ whose main role involves being small, infinite series involve a natural number n whose main role is to be large. Specifically, an infinite series

$$a_1 + a_2 + a_3 + \dots$$

converges to a limit L if, for any positive number ε there exists N such that

$$|(a_1 + a_2 + a_3 + \dots + a_n) - L| < \varepsilon$$

whenever $n > N$. In words: the sum of finitely many terms of the series becomes as close as we wish to L if the number of terms is large enough.

For example, we can define the recurring decimal $0\dot{3}$ to be the sum of the infinite series

$$\frac{3}{10} + \frac{3}{100} + \frac{3}{1000} + \frac{3}{10000} + \dots$$

The limit L of this series is *exactly* 1/3. The difference between L and the sum of the first n terms is

$$\frac{1}{3} - \left(\frac{3}{10} + \frac{3}{100} + \dots + \frac{3}{10^n} \right) = \frac{1}{3.10^n}$$

Given $\varepsilon > 0$, we can make the right-hand side less than ε by taking $N > 1/\varepsilon$, because $3.10^n > n$.

The same type of calculation proves that the series representing *any* infinite decimal converges to a real number, and that every real number can be expressed as a possibly infinite decimal. This justifies the use of infinite decimals as a conceptual notation for real numbers.

If no such L exists, the series *diverges*, and the infinite sum can't be given a meaning as a limit. Grandi's proof of Creation from nothing in Chapter 1 is fallacious because it uses a divergent series. However, some divergent series can be given a sensible meaning by inventing a new (technical) definition of the sum. Mathematicians eventually did that, for a class of divergent series said to be *summable*. This justifies Grandi's claim that

$$1 - 1 + 1 - 1 + 1 - 1 + \ldots = \tfrac{1}{2}$$

is a meaningful result in a specific technical context, but not his interpretation that something can be created from nothing.

Infinitesimals revenant

Limits resolved the paradoxical issues about infinity and infinitesimals in analysis, but the ghost of the infinitesimal didn't fade completely. In mathematics, it's unwise to abandon an interesting idea just because it's wrong. It's also unwise to keep pushing an incorrect idea without changing it or acknowledging the error, but incorrect ideas can sometimes be reformulated so that they work. Infinitesimals are a case in point.

When everyone thought that mathematical numbers were reality itself, albeit in an idealized form, *the* number system (basically, the real numbers) was the only one possible. (Complex numbers were accepted with reluctance at first, and then made respectable by thinking of them as pairs of real numbers.) So, if you defined an infinitesimal as 'a positive number that is smaller than any positive number' you were in trouble. It had to be smaller than itself.

If you stop thinking that the only possible numbers are the real numbers, however, there's a way out: define an infinitesimal as 'a positive number *of some novel kind* that is smaller than any positive *real* number'. Since it's not a real number, the argument that it must be smaller than itself fails. Making this idea work sensibly isn't straightforward, however. Once you throw in a new infinitesimal number—call it ε—then you have to make sense of all algebraic expressions involving ε, such as $\varepsilon + 5$, $11\varepsilon^2$, and $1/\varepsilon$.

To do analysis, you need to define sin ε, cos ε, log ε, e$^\varepsilon$, and so on. Then you have to prove that these extended concepts have all of the usual properties that we expect, and that the entire structure is logically consistent.

Assuming this can be done, your new number system also contains infinite numbers. By definition, $\varepsilon < 1/n$ for all natural numbers $n > 0$. Therefore $1/\varepsilon > n$ for all natural numbers $n > 0$, so $1/\varepsilon$ is infinite.

In 1877 Paul du Bois-Reymond began to develop just such a number system. In *Über die Paradoxen des Infinitär-Calcüls* (on the paradoxes of the infinitary calculus) he wrote:

> The infinitely small is a mathematical quantity and has all its properties in common with the finite ... Yet when one thinks boldly and freely, the initial distrust will soon mellow into a pleasant certainty ... A majority of educated people will admit an infinite in space and time, and not just an 'unboundedly large'. But they will only with difficulty believe in the infinitely small, despite the fact that the infinitely small has the same right to existence as the infinitely large.

Another pioneer of a meaningful notion of infinitesimal, at much the same time, was Otto Stolz. He extracted the key feature that excludes infinitesimals from the usual real numbers, naming it the Archimedean property because Archimedes stated it as an axiom when applying exhaustion in *On the Sphere and Cylinder*. The property concerned applies to any system with a sensible concept of 'less than' (that is, satisfying some reasonable axioms that I won't state here). It can be formulated as either of two equivalent statements:

- Every number x is less than some natural number n.
- If a number $x > 0$, then $1/n < x$ for some natural number n.

The first statement says that the system contains no infinite numbers, the second that it contains no infinitesimals.

Today we encapsulate this idea as: the real numbers are an Archimedean ordered field. That is, the usual operations of arithmetic can be defined and have the usual properties; a notion of 'less than' can be defined and has all the usual properties; finally, the Archimedean axiom applies. Indeed, **R** is the only Archimedean ordered field, except for trivial changes in notation ('up to isomorphism'). Stolz and du Bois-Reymond discovered that, in contrast, there exist many different non-Archimedean ordered fields. By definition these contain both infinite and infinitesimal 'numbers'. Du Bois-Reymond constructed a natural example in 1875: all real-valued functions of a real variable x, ordered by their 'asymptotic' behaviour for large x. The logarithmic function represents an infinitesimal element, and the exponential function represents an infinite element.

Non-standard analysis

Non-Archimedean ordered fields can be used to justify many features of analysis, such as the definition of the derivative in calculus, using genuine infinitesimals. But to do this systematically, we must define analogues of standard functions such as log, exp, sin, and so on. One way to do this is with power series expansions, but plenty of useful functions can't be defined by power series. So we have to establish exactly which properties of the real numbers have sensible analogues.

Much better would be to construct the field so that every important property of the real numbers (except those like 'being Archimedean') automatically has a sensible analogue. This is easier said than done, and no one expected it to be possible, but in the early 1960s Abraham Robinson discovered that it is. A key step was to distinguish properties that still work (such as $e^{a+b} = e^a e^b$) from those that don't ('the field is Archimedean'). His 1966 *Non-Standard Analysis* states:

> The idea of infinitely small or *infinitesimal* quantities seems to appeal naturally to our intuition...Leibniz argued that the theory of infinitesimals implies the introduction of ideal numbers which might be infinitely small or infinitely large compared with the real numbers but which were *to possess the same properties as the latter*...It is shown in this book that Leibniz's ideas can be fully vindicated and that they lead to a novel and fruitful approach to classical analysis and to many other branches of mathematics.

Robinson's discovery emerged from a branch of mathematical logic known as model theory, which examines the relation between systems of axioms and mathematical structures that satisfy them. Using model theory, he proved the existence of a non-Archimedean ordered field, having all the properties of the real numbers that can be expressed by logical statements that have 'bounded quantification'—a technical restriction on the use of the quantifiers 'there exists' and 'for all'. 'Being Archimedean' can't be expressed in that manner.

Any non-Archimedean ordered field of this type is known as 'the' *hyperreals*, denoted $\mathbf{R}^*$. Such fields are not unique, but any of them can be used as a model for properties of the real numbers with bounded quantification. Now we can prove the *transfer principle* that any bounded-quantification property is valid for $\mathbf{R}$ if and only if the analogous property is valid for $\mathbf{R}^*$. This implies, for example, that standard functions such as log, exp, sin, and cos can be defined in $\mathbf{R}^*$, so that all of their usual properties, such as $\log xy = \log x + \log y$, remain valid.

Since the set $\mathbf{R}^*$ of hyperreals forms a non-Archimedean ordered field, it contains infinitesimals, and their reciprocals are infinities. Every finite hyperreal can be decomposed uniquely as a real number plus an infinitesimal. The real number is called its standard part. Now classical limiting processes can be cast in the language of infinitesimals, provided we take the standard part at the end. The derivative of a function f can be defined not as a limit, but as

$$f'(x) = \text{st}\left(\frac{f(x+\varepsilon) - f(x)}{\varepsilon}\right)$$

where ε is infinitesimal and st is the standard part.

By such means, all uses of limits in analysis can be replaced by intuitive reasoning about infinities and infinitesimals, along lines that go right back to Newton, Leibniz, and indeed predecessors such as Fermat, Cavalieri, and Archimedes. The only missing ideas were the existence of hyperreals, the proof of the transfer principle, and—above all—*taking the standard part* in the usual formulas.

Some educators have argued that non-standard analysis provides an effective new way to introduce analysis to undergraduates, who often have difficulties with limits. The method has been tried out in university classes, with some success. One psychological barrier is that the definition of hyperreals isn't constructive. It doesn't provide a specific model for $\mathbf{R}^*$ in the way that $\mathbf{R}^2$ does for Euclidean geometry and infinite decimals do for the real numbers. Another is that the model-theoretic proofs are very abstract. So the classical limit approach to analysis is still the default for most mathematicians and most undergraduate lecture courses.

Important theorems have been proved using non-standard analysis, in areas such as probability theory and fluid dynamics. By the transfer principle, standard proofs of these theorems must exist, but they're not actually known. In practice the main impact of non-standard analysis has been a philosophical one: it provides a logical framework for infinitesimals.

Chapter 5
Geometric infinity

If you stand next to a long straight railway line, you get a strong impression that the railway tracks meet on the horizon (Figure 12). This happens because the tracks are parallel. The parallel edges of long straight roads behave in the same manner.

In the familiar geometry of Euclid, parallel lines play a special role. By definition, two lines are parallel if they're always the same distance apart, so they can't meet however far they're extended. However, if an observer stands on an infinite plane between two parallel lines, then the further away they get, the closer they seem to become. In some sense, the lines appear to meet 'at infinity'—a discovery that inspired one unknown schoolboy to state that 'infinity is where things happen that don't'.

He was right. Mathematicians found a way to extend Euclid's geometry, by adding an extra 'line at infinity' to represent the horizon. Ordinary lines are similarly extended by equipping each of them with an extra 'point at infinity'. This idea led to a new and extremely fruitful kind of geometry called projective geometry. Historically, a major source of motivation was the visual arts. The artists of the Italian Renaissance wanted to paint three-dimensional objects and scenes so that they looked realistic.

12. Parallel railway tracks appear to meet on the horizon.

Linear perspective

The apparent convergence of parallel railways lines is one of the simplest examples of *linear perspective*: the accurate representation of three-dimensional geometric forms on a flat, two-dimensional canvas. Distant objects appear to be smaller. You can cover the Moon with your thumb; the sheep across the field looks a lot shorter than the one staring at you over the fence. These familiar effects are a consequence of the physics of light rays and the structure of the human visual system. However, it's not enough to make sheep in the foreground larger than those in the background. The entire geometry of the painting has to fit together in a systematic way that represents how objects in three dimensions appear to the eye.

Before the Renaissance, artists either ignored this issue or got it wrong. Ancient Egyptian art, for example, ignores it: the size of a person in a relief is largely dictated by their social importance. Pharaohs and gods tower over ordinary mortals; servants are smaller than their masters; wives are often (but not always) smaller than their husbands. In the relief of Figure 13, Ramses II is depicted as being taller than his horses, and far bigger than anyone else. The artist's need to fit a lot of detail into a limited space is another factor. In particular, the fortification at the right is shown in a stylized form, not in perspective. (These remarks aren't intended as criticism: the Egyptians made a deliberate stylistic choice. Other reliefs are remarkably realistic, especially depictions of birds and other living creatures.) Medieval artists sometimes tried to depict buildings in rudimentary perspective, but failed to relate them to each other in a coherent way.

Renaissance artists benefited from a deeper understanding of the geometry of perspective drawing. The earliest systematic method was probably that of Ambrogio Lorenzetti, employed in his 1344 *Annunciation*, which includes a very accurate tiled floor. Filippo Brunelleschi took a big step forward in the early 15th century with images of the Florentine Baptistery and the Palazzo Vecchio. A major advance came with Leon Battista Alberti's *Della Pittura*, begun in 1435. The basic ideas arose from a combination of artistic imperatives, experimental observations, and geometric reasoning.

Figure 14 shows one of the most famous of the early perspective paintings, Piero della Francesca's *Flagellation of Christ*. The artist flaunts his mastery of perspective to considerable effect. The interpretation of many features, notably the large figures in the foreground, is controversial, but the tiled floor, with its chequered pattern, is impressively accurate yet understated, as are the buildings.

The techniques of perspective came from an idealization of the geometry of human vision that doesn't correspond exactly to reality. However, this idealization has considerable mathematical interest. It was a huge improvement on previous practice in the visual arts and was an early step towards the scientific

13. Ramses II's victory at the Siege of Dapur. From a relief in his temple at Thebes.

14. *Flagellation of Christ* by Piero della Francesca.

understanding of vision. These artistic endeavours also had a significant effect on mathematics, motivating a new kind of geometry. It's called projective geometry, because it centres on how images can be projected from one plane to another, much as a frame of a movie is displayed on a cinema screen.

In the abstract, this process sets up a transformation from one plane to another. Choose a fixed but arbitrary point not lying on either plane, the centre of projection. Given a point on one plane, draw the straight line through that point and the centre of projection. This meets the other plane in a unique point, the image of the first under the projection. In Figure 15, Albrecht Dürer is experimenting with a projection from the plane of a table, supporting a lute, to a vertical screen that has helpfully been hinged open to show the resulting image.

Anyone using a video projector runs into one annoying feature of this technique: 'keystoning', in which a rectangular picture is distorted so that it looks like the keystone at the top of an

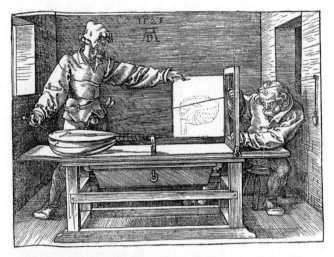

15. Albrecht Dürer, from *Underweysung der Messung Mit dem Zirckel un Richtscheyt, in Linien*, Nuremberg, **1525**.

arch, wider at the top than at the bottom. Video projectors usually have control settings to eliminate this effect. Instead of trying to get rid of distortions introduced by projection, projective geometry revels in them. Its core objective is to find geometric properties that are not affected by such distortions, just as much of Euclidean geometry is really about features of shapes, such as lengths and angles, that remain unchanged after rigid motions.

Beyond the blue horizon

One of the simplest proofs that the Earth isn't flat is the existence of a horizon. As a ship travels further away from harbour, it begins to disappear from view where the ocean meets the sky. First, the lowest portions are obscured, then higher ones, until eventually only the tip of the mast can be seen. Then that, too, dips below the horizon, and the ship has vanished entirely.

Reality is more complex: atmospheric effects can distort the paths taken by light rays between ship and eye, variations in heat can create mirages, haze can cause the ship to fade slowly from view, and the surface of the ocean goes up and down as waves pass across the line of sight. Ignoring these subtleties, Figure 16 shows a circular Earth in the plane, a cross-section of a spherical Earth in space. The vanishing act is governed by a straight line from the observer's eye, tangent to the surface of the ocean. As the curvature of the Earth takes the ship below this line, the lower parts of the ship cease to be visible, obscured by the ocean. The point of contact of the tangent line with the surface of the Earth is the horizon. In three dimensions, on a perfectly spherical Earth, and observing in all directions from the same point, the horizon forms a circle centred on the observer's eye.

The Earth is finite, but something similar occurs if our spherical planet is replaced by a Euclidean plane of infinite extent. The ship no longer disappears, but there's still a well-defined horizon. In cross-section, the direction to the horizon is shown in Figure 17 as a solid arrow parallel to the ocean surface.

The dotted lines are lines of sight to the top of the mainmast. The further away the ship is, the closer this line gets to the solid one. So the top of the ship appears to rise towards the direction of the

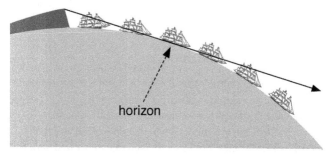

horizon

16. A ship disappearing over the horizon.

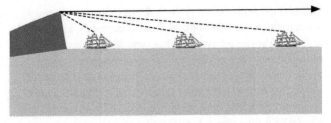

17. A ship on a planar ocean, relative to the direction of the horizon.

18. As the ship moves further away, it rises towards the horizon (dotted line) and appears to become smaller.

horizon. The bottom of the ship (line of sight not shown) also rises, and the angle between these two lines shrinks, so the ship becomes smaller and smaller to the eye as its angular height decreases (Figure 18).

If the observer's line of sight is above the solid line, it fails to meet the surface of the ocean (although its backward extension would meet it, behind the observer, if there were ocean in that direction and nothing obscured it). If it's below this line, it meets the ocean at some specific point. The line itself, being parallel to the ocean surface, doesn't meet it, but demarcates the *boundary* between directions that meet the ocean and those that don't.

This boundary is shown dotted in Figure 18. Every direction below the boundary terminates at some point of the plane. However, the boundary itself does *not* correspond to points on the Euclidean plane defining the ocean. The line in the direction of the horizon is parallel to the plane, so can't meet it.

77

This example is also about projection, this time from the real world to the retina of the observer (more accurately, a plane held in front of the retina). We have thus come to the paradoxical conclusion that in a geometrically accurate projected image of a plane, there exists a line that is not the image of any line on the plane. This curious fact remains true even if we allow lines to be extended backwards. Now points above the horizon correspond to points of the plane behind the observer, points below the horizon correspond to points of the plane in front of the observer, but points on the horizon don't correspond to any point of the plane.

Nevertheless, the horizon is a uniquely defined straight line in the image, with a specific geometric meaning.

A similar effect arises for a straight railway line. To the observer, the lines seem to meet on the horizon. This isn't literally true on our round planet—parallel lines don't exist on a sphere—but it is exactly true of a plane in Euclid's geometry. There seems to be a sense in which parallel lines *do* meet, but not at any point in the plane. Projective geometry handles this effect by adding an extra 'line at infinity' to the Euclidean plane. But it took a while for this point of view to emerge, and even longer to replace it by something better.

The line at infinity

Mathematicians dislike exceptions to otherwise general rules. Parallel lines are an example: any two distinct lines meet at exactly one point... oops, sorry: except if they're parallel. That's especially ugly given that any two distinct points can be joined by exactly one line. 'Parallel points' that can't be joined don't exist. Why are points well behaved, whereas lines behave badly?

It took many centuries before the mathematical world realized that exceptions can often be removed by artificially extending the underlying structure. Minus one has no square root: no worries, just define a new kind of number to supply one. Prime factorization is not unique for algebraic numbers: just define a new kind of number so that it is. History is full of words like 'imaginary' and 'ideal', associated with this procedure, but nowadays throwing in new ingredients to fix up an exception is considered fair game, just as sensible as anything else in mathematics. So it's natural to upgrade the Euclidean plane by plugging in a *line at infinity*, and to deem parallel lines to meet there.

It took time for mathematicians to feel comfortable with such processes, which initially were rather mysterious. Eventually, a suitable logical framework emerged, within which the constructions made sense *and* it was possible to prove that they worked.

To illustrate how difficult these ideas seemed before their logical status was clarified, I'm going to introduce the line at infinity in an intuitive manner first, and clarify the mathematics—and the meaning of infinity in this context—afterwards. If this makes some of the arguments confusing and mysterious, don't worry: I'm just putting you in the shoes of the mathematicians who had to grapple with similar confusions.

The horizon is a good example of an annoying exception. The real horizon exists, because ships can sail over it. But that's because the Earth is curved. If the Earth, and its ocean, were an infinite flat plane, the artist would still see a well-defined horizon, even though no ship could ever reach it. It's as if Euclid's plane has a line missing, whereas the image does not. So mathematicians decided to give the plane an extra line to plug the gap, and they called it the line at infinity. Intuitively, it's what the boundary of the plane would be if it *had* a boundary. Head off in any given direction, walk forever, and when you've arrived, you've reached the line at infinity.

Figure 19 is a schematic of this idea. The entire Euclidean plane is squashed inside a circular disc. The plane is the interior of the

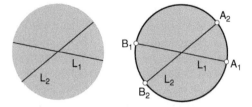

19. *Left*: representation of the Euclidean plane as a disc (shaded) without boundary, and two straight lines through the origin. *Right*: adding an extra line at infinity (dark circle). The extra points (white dots) at the end of each line are identified in diametrically opposite pairs, so that lines meet at only one point.

disc, and the line at infinity is its boundary; that is, the circle itself, as in the right-hand figure. A line L_1 in the Euclidean plane fails to meet the circle, because the boundary is not part of the Euclidean plane. (The endpoints of the line segments in the left-hand figure are not part of L_1 and L_2.) However, if L_1 is extended in the natural way, by adding points A_1 and B_1, it meets the line at infinity at those points. Start at the origin, head off along L_1, and eventually, after walking an infinite distance, you get to A_1, on the line at infinity. Walk the other way and you get to B_1. We obtain an extension of the line lying in an extension of the plane—both beyond the scope of Euclid.

In the Euclidean plane, the lines L_1 and L_2 meet at only one point: the origin. No extra point at infinity lies on both of them, so they don't meet each other at infinity either. That's good, because they're not parallel. (I'll come to parallel lines, and where they meet, shortly.) However, taking the figure literally, L_1 meets the line at infinity in *two* points A_1 and B_1, one at each end. The same goes for L_1. As well as being inelegant, this property makes the line at infinity exceptional, which is exactly what we were trying to avoid when we introduced it.

To get round that, we're forced to do something that at first sight seems very strange. We must *identify* the two points A_1 and B_1.

The same goes for any other line in the Euclidean plane. That is, we must redefine the concept of point—for these extra points at infinity only—as 'pair of diametrically opposite points'. This means that every point on the line at infinity seems to appear twice in the picture. As a consequence, so does the line at infinity itself. Start at A_1 and walk clockwise along the line at infinity until you get to B_1. Now you're back at the starting point, because B_1 is the same as A_1 after identification. If you continue walking clockwise, you cover the same ground a second time, until you get back to A_1. So the line at infinity isn't really a circle; it's a semicircle. Except that the two ends of the semicircle, A_1 and B_1, are the same point, so the semicircle 'wraps round' and its ends join up. So actually it *is* a circle, topologically speaking. Just not the one that is most evident to the eye.

This may seem to make the line at infinity special, something I've been trying desperately to avoid. But now that we have those extra points at infinity, Euclidean lines also wrap round. Start at the origin, head off along L_1 until you get to infinity; that is, A_1. You need not stop there, because A_1 is the same as B_1, and from B_1 you can continue along L_1 to get back to the origin. So L_1, plus its new point at infinity, also closes up on itself to form a topological circle. Democracy prevails among the lines. In fact, any line, Euclidean or at infinity, can be projected to any other line, while preserving all geometric features.

The line at infinity is very weird: it's circular and straight at the same time. It's straight in the sense that in whichever direction the artist looks, the horizon in the image is straight. But it's a circle in the sense that, if the artist slowly spins round through 360 degrees, that perfectly straight horizon keeps extending and extending...until eventually it joins up with itself. Ordinary lines in Euclid's geometry don't do that. They head off forever along one direction, and they head off forever along the exactly opposite direction. The longer you make the line, the further apart its endpoints become. They certainly don't start to close up.

However, when we add the line at infinity to the plane, we create a knock-on effect on ordinary lines. Any ordinary line in the plane is transformed into a line in the image, which typically meets the horizon. Since this 'intersection point' lies on the horizon, it doesn't correspond to anything on the plane. Although the image of an ordinary line seems to cross the horizon, there's no point *on that line* whose image is the crossing-point. Nevertheless, the crossing-point does correspond to a point on the line at infinity. It arises when we equip an ordinary line on the plane with an extra point at infinity.

The geometry of perspective drawing makes sense of my apparently arbitrary decision to identify the two 'ends' of a Euclidean line. As a finite line is made ever longer, the images of its endpoints both approach the horizon from different sides. Figure 20 shows an artist, drawn as an eye on a stick, standing on the plane (shaded). The dashed line with arrows is parallel to the plane, and also to the line in the plane that runs through points P_1 and P_2, one in front of the artist, the other behind. The artist's view of the horizon is the dotted line on the canvas (white parallelogram). The images of P_1 and P_2 on the canvas are Q_1,

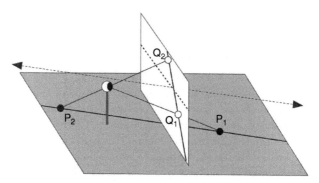

20. **How the artist's eye projects a line in the plane.**

below the horizon, and Q_2, above it. The difference is that the artist's eye is between P_2 and Q_2, but not between P_1 and Q_1.

What we're doing here is making mathematical sense of the intuition that a circle of infinite radius is a straight line. However, we've modified intuition slightly to get something sensible. Our circle of infinite radius is actually a straight line *plus an extra point*. That closes it up to form a circle.

The weirdness doesn't stop there. If two circles meet, they generally do so at two points (unless they're mutually tangent). So shouldn't an ordinary line, suitably augmented by a point at infinity, meet the line at infinity in *two* points? The answer is 'no', and the reason is that although an ever-expanding line has two ends, their images both approach exactly the same point on the horizon.

Parallels

To see what happens to parallel lines, I need to specify how to represent the entire plane as the interior of a circle, to make sense of Figure 19. One way—there are others—is shown in Figure 21. Here a hemisphere with centre C sits so that it touches the plane at one point, say the origin. Its vertical projection is a disc, shown in pale shading. Take any point P in the plane and project it to Q on the hemisphere. Then project Q vertically to R. Now every point P in the plane is mapped to a point R interior to the circle, because the equator of the sphere lies in a parallel plane. The boundary of the hemisphere, projected down to the plane, becomes the boundary of the circle.

We now have a recipe that turns geometry on the plane into geometry on the interior of the circle. Just transform each point of the plane according to the procedure just outlined. A line in the plane, through the origin, transforms into a semicircle on the hemisphere, which is half of a great circle. This projects back

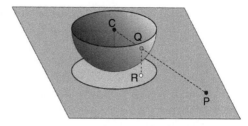

21. How to fit an infinite plane inside a circle. Project point P to Q on a hemisphere, centre C. Then project Q vertically to R.

down to form a radial line *segment*, a diameter of the disc. As the original line tends to infinity, its image approaches the boundary circle. It approaches the boundary circle at two such points, one for each direction along the line. In the projective plane, these two points are identified. So now we have a geometric model for the projective plane, obtained using Euclidean three-dimensional geometry as an intermediary.

A line not passing through the origin also transforms into a great semicircle on the hemisphere, because it lies on the intersection of the hemisphere with the plane through the line and C. Its vertical projection is half an ellipse.

We can now find out how parallel lines behave. Figure 22 (left) shows five parallel lines in the Euclidean plane, transformed as just described. Although they appear, to the eye, to meet on the boundary of the disc, the boundary corresponds to the line at infinity, not to anything in the Euclidean plane. When the line at infinity is adjoined, as in Figure 22 (right), all five lines meet each other, and the line at infinity, in a single point—the same one at either end, despite appearances, because of the identification rule.

I said that in Euclidean geometry any two lines meet at exactly one point; adding 'oops, sorry: except if they're parallel'. But in our

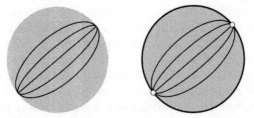

22. Parallel lines in the Euclidean plane (left) meet at infinity (right).

spanking new geometry with an extra line at infinity, there's no 'oops'. *Any* two lines meet in a single point. Yes, it may be at infinity, but let's not be picky. That's far more elegant than making an exception. Parallel lines fit into this picture beautifully. Their images both pass through the *same* point on the horizon, so their extended versions both pass through the same point at infinity. In short: *parallel lines meet at infinity*. The schoolboy was right.

Directions

Today we approach the whole topic in a different way. The key concept is not new fictitious points on a plane, but existing directions in space. If you look back at the discussion, the idea of the artist looking in a given direction, or parallel lines all pointing in the same direction, keeps coming up. This is a clue. But what is a direction? The easy way to derive a good model from ordinary geometry is to start in three-dimensional space, with a distinguished point, the origin. That's where the artist conceptually places her all-seeing eye. A direction is then determined by a line that passes through the origin. Any line parallel to that one 'points in the same direction'.

The end result of about 600 years of effort (plus some even earlier work that we now see fits into the same framework) dispels the mystery surrounding points at infinity. The entire set-up can be interpreted in terms of standard Euclidean geometry in three dimensions, but taking 'points' to be lines through the origin, not Euclidean points. Reducing the dimension by intersecting with a sphere centred at the origin, projective points become diametrically opposite pairs of Euclidean points. The hemisphere in Figure 22 is an optional extra; making it clear how the Euclidean plane embeds in the projective plane. And the line at infinity is no longer a fictitious object introduced to represent what an artist sees: it's the equator of a finite object, the sphere.

Perspective drawing

I'm not going to try to teach you perspective, but you ought to be shown some payoff in the visual arts. The line at infinity yields a simple solution to what otherwise would be a very complicated problem in geometry. Renaissance painters, and many later ones, made extensive use of floors decorated with a regular grid of square tiles. Johannes Vermeer even used them when depicting buildings that didn't actually have them, and he wasn't alone. One reason may be that tiled patterns are visually impressive, and very hard to get right without projective geometry. Another is that they create a grid, giving visual clues about depth.

The main ideas can be seen in a simple instance: drawing a 4 × 4 pattern of square tiles in perspective. Figure 23 shows the

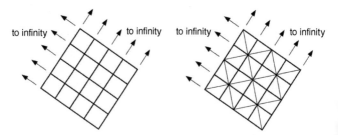

23. *Left*: a grid of squares with two directions 'pointing to infinity' corresponding to sets of parallel lines. *Right*: auxiliary diagonal lines used to construct a perspective version.

pattern, together with some arrows that point 'to infinity'. Even though the line at infinity doesn't exist in Euclid's geometry, we'll use the corresponding points on it to draw a perspective version of the grid.

The stages of construction (Figure 24) go like this:

1. Draw a horizon (dotted line) corresponding to the line at infinity. Draw three corners (black dots) and the two lower sides of the grid (thick lines). The dots can be at any location, other than having all three in a straight line, because any set of three points can be projected to any other set of three points. This choice corresponds to looking down on the grid from a slight angle.

2. Continue the solid lines to meet the line at infinity (shaded dots).

3. To draw the other two edges of the grid, observe that they're parallel to their opposite edges and so meet those edges on the line at infinity, at the points already constructed—one for each set of parallels. The intersection of these lines gives the fourth corner of the grid (black dot at top), and the other two sides can be filled in (solid lines).

4. Draw the diagonals of the square and find the point where they cross (white dot).

5. In the original grid, two grid lines pass through the centre of the square (where its diagonals meet). In the image, the corresponding lines do the same. Since they're parallel to appropriate edges of the square, they also pass through the same point at infinity, which determines where to draw them.

6. Subdivide the four squares of the grid now constructed by drawing their diagonals and finding their intersections: compare Figure 24 (bottom). The remaining grid lines pass through these intersections and the appropriate point at infinity (grey dot) for the same reason as in step 5.

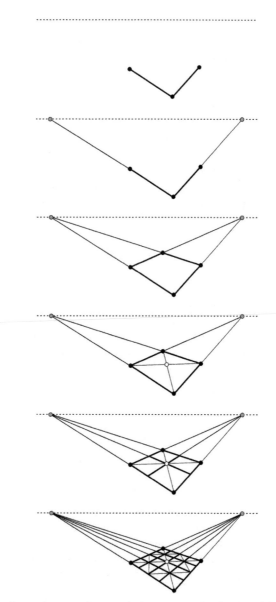

Infinity

24. Successive stages in constructing a perspective drawing of the grid.

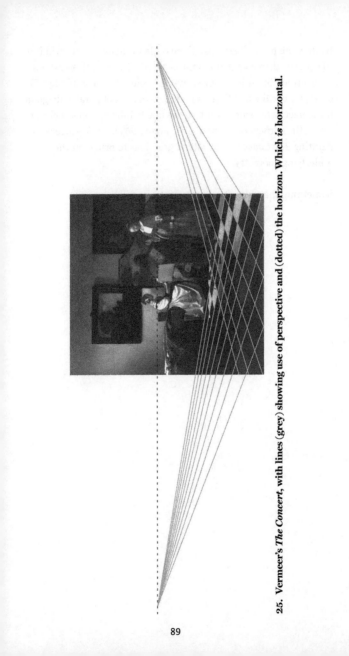

25. Vermeer's *The Concert*, with lines (grey) showing use of perspective and (dotted) the horizon. Which *is* horizontal.

By drawing more diagonals, it's possible to subdivide the grid into 16 squares, then 64 squares, and so on. It's also possible to extend such a tiling by adding more rows and columns along its edge. To do so, we use the fact that all members of a set of parallel diagonal lines meet at the same point on the line at infinity. These help to locate the corners of the new tiles. Figure 25 shows Vermeer's painting *The Concert*, with added grid lines to bring out the underlying geometry.

Something that doesn't actually exist can still be useful.

Chapter 6
Physical infinity

Infinity is virtually indispensable in today's mathematics, but there it's a conceptual entity, not a real one. Philosophers think about the infinite, and argue about whether it exists, and if so, in what sense. Religions often claim infinity as an attribute of their god or gods, and people have been executed for denying it, but today it's generally agreed that the existence of a deity is a matter of faith, not objective evidence.

Does infinity exist outside the human mind? Can it be *real*, not just in the sense of religious faith, but in the sense that rivers, trees, cats, and rocks exist? No philosophical hair-splitting about what 'real' and 'exist' mean: can someone *show* us infinity?

In virtually every area of the physical sciences, infinity is an embarrassment. A theory that predicts infinities is wrong. That doesn't mean it's useless, but it needs tweaking to get rid of those pesky infinities. However, there's one area of physics in which an actual infinity—physical, not conceptual—is not just tolerated, but presented as a possible truth: cosmology.

I'll start with more mundane occurrences of the infinite in theoretical physics. Here, infinite quantities are usually referred to as *singularities*, and their presence is evidence of defects in the model. However, it may still be very accurate away from singularities. I'll discuss singularities in three physical contexts: optics, Newtonian gravity, and Albert Einstein's relativity. Then we'll take a quick look at whether the universe is infinite.

Infinity in optics

Many of nature's glorious spectacles are created by unusual effects involving light. The rainbow is the most familiar, a narrow multicoloured arc across the sky. Another is the glory, in which a person gazing into mist with the Sun behind them sees a rainbow-like halo round the shadow of their own head. Someone standing next to them sees much the same thing, but now the halo is round *their* head. The glory may have inspired the tradition of depicting holy figures with halos, which goes back at least to 1st-century Buddhism. Complete circular halos sometimes surround the Sun or the Moon, with rarer effects such as light pillars (a vertical column of light above the rising or setting Sun) and sun dogs (bright spots either side of the Sun).

These effects happen when light from the Sun is reflected and/or refracted—bent—by water droplets or ice crystals. A rainbow appears when the Sun is behind the observer and rain falling from clouds is in front of them. Sunlight hits each drop of rain and is refracted; then it reflects off the back of the droplet; finally it is refracted on the way out (Figure 26). The angular radius of the rainbow is about 42·5°, and its colours are ordered by wavelength. A dimmer rainbow, outside the main one, is often visible, caused by a further reflection and refraction; its angular radius is about 52°. The glory is similarly created by sunlight passing through drops of water in mist, but for complicated physical reasons it almost reverses its path, which is why it appears to emanate from the shadow of the observer's head. It consists of a complex series of coloured rings of differing brightness. Halos, both solar and lunar, are caused by ice crystals in the upper atmosphere. The simplest form is a circular ring with angular radius about 22°, which is related to the geometry of ice crystals.

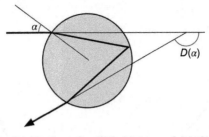

26. Deviation $D(\alpha)$ in the angle of light hitting a spherical water droplet.

All these phenomena are multicoloured because light of different wavelengths produces different colours, and the Sun's white light is a mixture of colours; indeed, 'all the colours of the rainbow', along with other wavelengths the human eye can't detect. The rainbow for light of a single wavelength is an arc of a circle, centred at a point diametrically opposite the Sun, and its angular radius is the rainbow angle for that wavelength. Different wavelengths have different rainbow angles in a continuous range, whence the band of concentric arcs of different colours that we see in the sky.

'Wavelength' is a concept in wave optics, where light is considered to be a wave. But the first serious mathematical description of light was ray optics, in which light travels along straight lines in any medium with constant refractive index. When we use ray optics to calculate the intensity of light at the rainbow angle for light of a single wavelength, the answer is 'infinity'. Let's see why.

Figure 27 (top) shows how the deviation $D(\alpha)$ varies with α. For a specific shade of red light, this curve has a minimum at $\alpha = 59 \cdot 6°$, and the corresponding deviation is $137 \cdot 5°$. Call this the rainbow angle. It exceeds 90° because the ray reverses direction, so the angular radius of the arc is the difference $180° - 137 \cdot 5°$, which is $42 \cdot 5°$. Figure 27 (bottom) shows what happens to an incoming band of light rays. They emerge at a range of angles, shown by the

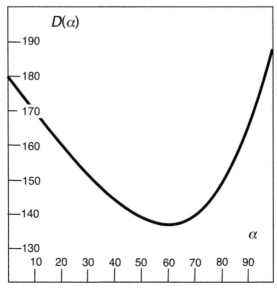

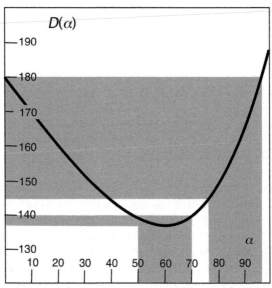

27. *Top*: graph of $D(\alpha)$ against α. *Bottom*: incoming light is compressed near the rainbow angle.

shaded regions. If the band enters near the minimum of the graph, it's compressed into a very small range of outgoing angles. So the outgoing light is concentrated in almost the same direction, making it brighter. If the incoming rays are very close to the minimum, the intensity of the emerging light becomes arbitrarily large, proportional to the reciprocal of the slope of the curve. At the rainbow angle, the slope is zero and the theoretical intensity is 1/0: infinite.

This calculation is far from useless. It gives the rainbow angle very accurately, and it goes some way towards explaining why the light at the rainbow angle is much brighter than at any other angle. That's why we see a sharply defined arc. On the other hand, the prediction of infinite intensity can't be the literal truth. So ray optics predicts a singularity in intensity, which physicists have learned, with good reason, to reject.

Ray optical calculations break down near the rainbow angle, a *geometric* singularity. The problem was resolved by the discovery that light travels not as a ray, but as a wave. The intensity around the rainbow angle could then be recalculated. It turned out to be something called an Airy function, which oscillates, and is large but finite at the rainbow angle (Figure 28). The oscillations represent diffraction fringes, a wave-optical effect that gives parallel light and dark stripes.

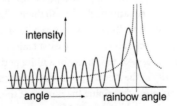

28. **Airy function of wave optics (solid line) and ray-optical intensity (dashed) near the rainbow angle. The intensity calculated using wave optics is finite. Its peak is near, but not equal to, the ray-optical rainbow angle.**

Infinity in Newtonian gravity

A more dramatic example of infinity occurs in Newtonian gravitation. Recall Newton's Law of Gravity: any two bodies in the universe attract each other with a force proportional to their masses and inversely proportional to the square of the distance between them. Applied to two bodies, the law predicts that they orbit in ellipses about their common centre of mass. For three or more bodies, no simple geometric curve or formula exists; indeed, the orbits are often chaotic.

Around 1900, Henri Poincaré and Paul Painlevé asked: which singularities can arise in a system of particles obeying Newtonian gravity? That is, when do solutions cease to exist after some finite time? An obvious singularity occurs when particles collide, so the main issue was whether anything else can happen. In 1895 Painlevé proved that for three bodies, all singularities are collisions. He then asked what happens with four or more bodies. In 1908 Edvard von Zeipel upped the ante by proving that if non-collision singularities occur, particles become infinitely distant in finite time. This seemed so weird that mathematicians stopped thinking about the problem for the next fifty years.

Donald Saari took it up again in 1967, and by 1973 he had proved that non-collision singularities must also involve particles oscillating arbitrarily fast. Particles must approach other distant particles arbitrarily often, and get arbitrarily close to them. That may seem even weirder, but in 1988 Zhihong Xia proved it can happen for five or more bodies. In consequence, there exist trajectories such that *all* of the bodies disappear to infinity in a finite period of time. This is a singularity in the Newtonian model: mathematically, the solution to the equation describing the motion of the bodies can't be continued past some particular finite time.

In Xia's set-up (Figure 29), two pairs of bodies orbit tightly round each other in highly elliptical orbits, in planes at right angles to

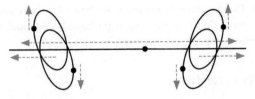

29. Xia's scenario.

a line along which the fifth particle moves. The pairs play celestial tennis with the fifth particle, which shuttles back and forth between them at an ever-faster rate. The pairs move away from each other along the straight line, and since energy, momentum, and angular momentum must be conserved, the bodies in each pair get closer and closer together. Everything speeds up so rapidly that after a finite period of time, all five particles disappear to infinity.

Later, similar behaviour was found in two-dimensional space as well.

Clearly this behaviour isn't physical. In a relativistic model, it can't happen because nothing can exceed the speed of light, but that's not the only issue. In a Newtonian model, there's no cosmic speed limit. So where does the model fail to match reality? The answer is: the use of point particles. No real body is a mathematical point. Usually, though, this simplification does little harm. Newton proved that the gravitational field outside a spherically symmetric body is the same as it would be if all of the mass were concentrated at the central point, so for many practical purposes a planet can be modelled as a point mass.

Sometimes, though, this assumption makes no sense. With bodies of non-zero size, Xia's scenario can't happen. When they get close enough, they touch. So the five-body disappearing act is an artefact of the assumption of non-physical point particles. Indeed, the gravitational potential of a point mass is inversely proportional to distance, so the potential tends to infinity as the distance tends to zero. The singularity in the five-body dynamics is a consequence of this singularity in the Newtonian gravitational potential. It's an artefact of the mathematical model.

Infinity in relativity

In special relativity, singularities occur when matter travels at the speed of light. Time grinds to a halt, lengths shrink to zero, and mass becomes infinite. However, these effects don't correspond to a physical singularity because no real body can attain the speed of light. The energy needed to accelerate it to that speed would itself be infinite. Of course, light travels at the speed of light, but a proper treatment of light requires quantum mechanics.

General relativity involves a more intriguing singularity. Einstein introduced general relativity to include gravity in his theories of space, time, and matter. In Newtonian physics, gravity is a force, acting between any two bodies. Newton didn't specify how a force can act across empty space. He was aware of the philosophical problem of action at a distance, but he took a pragmatic view and ignored it. Einstein replaced this force by the curvature of space-time. A planet orbiting a star follows a curved orbit not because of an attracting force, but because the star warps space and the planet is affected by the warp.

General relativity explains many gravitational effects that are inconsistent with Newtonian physics, such as a slow change in the position of Mercury in its orbit when it's nearest the Sun. GPS satnav systems process their timing signals using general relativity, because they'd give the wrong position if they didn't compensate for time-warping by the Earth's gravitational field.

One of the surprising consequences of general relativity is the existence of black holes. When a very massive star contracts under its own gravity, it can become so dense that its theoretical escape velocity—the initial velocity needed to propel a body fast enough to get away—exceeds the speed of light. Since that's impossible, so is escape. Even light remains trapped. Evidence for the existence of black holes has been accumulating steadily, and while there are some serious differences of opinion about quantum-mechanical fine points, astronomers generally accept that something very like the predicted structure exists.

The simplest kind of black hole, a static one, is surrounded by a spherical shell, its event horizon. Bodies can escape if they're outside the event horizon, but not if they're inside. It's therefore impossible for an external observer to find out what happens inside a black hole. Theoretically, a collapsing star should continue collapsing until, after a finite time, all of its mass is concentrated at a single point of infinite density. This, if it actually happened, would be a genuine physical singularity. Most physicists think that something else happens, avoiding a true singularity, but they're divided on what that might be. If the black hole rotates, the singularity becomes a circle, but the density of the matter from the star is still infinite.

Infinite universe?

Virtually the only context in which scientists don't consider an infinite quantity to be a sign that their theories are wrong is cosmology. At various times it has been entirely respectable to assert that the age of the universe, or its size, is infinite.

The simplest way to explain the origin of the universe—both in space and time—is to maintain that it never had one. If the universe has always existed, we can stop worrying about how it came into being. It's a seductive line of reasoning, although Aquinas wouldn't have approved, since he rejected the absence of a First Cause. But it was good enough for Fred Hoyle, forming the basis of his steady state theory, in which an infinitely large and infinitely ancient universe continually expands by the gradual creation of new particle–antiparticle pairs in the dark between the stars. Hoyle's theory was widely accepted in the 1950s, but by the 1970s most physicists and cosmologists had abandoned it in favour of the Big Bang: both space and time came into existence from a point singularity 13·8 billion years ago.

As Aquinas argued, this type of explanation isn't entirely satisfactory. It explains the universe away, rather than explaining it, by begging a simple supplementary question: *why* has it always existed? There's no mathematical difficulty in contemplating an infinitely old universe and still asking 'where did all of that come from?' Pushing a problem away to minus infinity doesn't entirely get rid of it, and in any case the current consensus among astronomers and cosmologists is that the universe hasn't always existed. However, it took a while for the Big Bang to gain acceptance. Hoyle gave it that name to poke fun at it. We had trouble getting it out of our heads that the universe has always existed—that its past is temporally infinite. This view keeps trying to sneak back, with questions like 'what happened before the Big Bang?'

We have even more trouble getting it out of our heads that the spatial universe must go on forever. Indeed, some cosmologists are convinced that it does. It might seem that when we gaze heavenwards on a starry night, we're looking into the infinite. There's no obvious boundary; the universe extends away from us for vast distances. However, there's a definite limit to the size of the observable universe. Currently, this is considered to be a sphere, centred by definition on the Earth, of radius 46·6 billion light years, or 44×10^{26} metres.

This figure appears paradoxical, because the universe has existed for a mere 13·8 billion years, and light travels at one light year per year. Since $46\cdot6/13\cdot8 = 3\cdot37$, the light now reaching us from the edge of the observable universe seems to have travelled at just over three times the speed of light, so the speed of light is three times itself. The paradox is resolved when we remember that the universe is expanding. When the light now reaching us from the edge of the observable universe first started out, the region that is now observable was much smaller. Calculations indicate it was only 42 *million* light years in radius.

We can also test the theoretical figure experimentally. In 2003 Neil Cornish and co-workers used observations of the cosmic microwave background (CMB) by the Wilkinson Microwave Anisotropy Probe to deduce that if the universe is finite, its diameter is at least 78 billion light years. They assumed that the universe has a closed topology, without any boundary, 'wrapping round' on itself like a sphere or a torus. Wrapping round creates pairs of circles in the sky, diametrically opposite each other, corresponding to light from the same region of the universe reaching us from two different directions (round the front and round the back, so to speak). These circles can be detected because the temperature pattern of the CMB is the same on both. They found no circles larger than 25° in angular radius with strongly correlated temperature patterns. This leads to the 78 billion light-year lower bound. In 2012 Cornish and others extended the estimate to 92 billion light years.

Curvature

The curvature of the universe is often proposed as a way to distinguish finite from infinite, on the reasonable assumption that the large-scale curvature is the same everywhere. Mathematicians recognize three different types of constant-curvature geometry: Euclidean, elliptic, and hyperbolic. Zero curvature gives Euclidean geometry. Positive curvature, like the surface of a sphere, gives elliptic geometry, so the elliptic analogue of the plane is finite. Negative curvature, like a saddle or a mountain pass, gives hyperbolic geometry, and the hyperbolic analogue of the plane is infinite. So it seems as though we could use curvature to distinguish finite space from infinite. If the curvature is positive, it's finite; if zero or negative, it's infinite.

You can still find this argument presented today, but unfortunately it's false. Cosmologists repeated the mistake from the 1930s to the 1990s. The correct statement is that there are three distinct space-time *metrics* of constant curvature, corresponding to the three geometries. But spaces with different topologies can have the same metric, and some of them can be finite when the corresponding geometry isn't. If you take Euclid's plane and roll it up into a cylinder, the metric doesn't change. Geometry on the cylinder is the same as on the plane, unless diagrams get so big that they wrap round and overlap themselves.

This is still an infinite space, but a similar trick leads to a finite one called a flat torus. The easiest way to visualize it is to start with a square, and then identify opposite edges—add a mathematical rule that they must be considered to be the same. The flat torus has the same metric as Euclidean space, but different topology, and it's finite. So zero curvature does *not* imply infinite size. Ironically, Karl Schwarzschild, whose work on general relativity contributed to the discovery of black holes, pointed this out in 1900. Mathematicians have also found similar constructions in hyperbolic geometry; instead of a square you use special polygons, and the range of possibilities is far richer. Aleksandr Friedmann told cosmologists not to assume negatively curved spaces must be infinite in 1924, but few listened. Despite what it says on many websites, these mathematical spaces show that you *can't* distinguish a finite universe from an infinite one by measuring its curvature.

Even if the universe really is infinite, it's difficult to see how this can be verified scientifically, since anything further away than 46·6 billion light years is unobservable. Of course there might be some way to infer that it's infinite without observing it, just as we can infer the temperature at the centre of the Sun without going there. But even then, it's hard to see how we could distinguish 'genuinely infinite' from 'very, very big'.

Chapter 7
Counting infinity

To my mind the most profound discovery about the infinite, one that all the philosophers (and all previous mathematicians, for that matter) missed, was made by Cantor in 1874. He demonstrated, logically and rigorously, that even within the realm of numbers, infinity comes in different sizes. Specifically, the infinitude of all real numbers is greater than that of the natural numbers 1, 2, 3,…. By this he didn't merely mean that some real numbers are not natural numbers, which is true but obvious. His proof showed that it's impossible to match every real number to a corresponding natural number, in such a way that distinct real numbers correspond to distinct natural numbers. The natural numbers are countably infinite, but the real numbers are uncountably infinite.

Cantor returned to this theorem in 1891, giving a different proof, his celebrated 'diagonal argument' described later. But in 1874 he was after more classical game. Mathematicians had distinguished *algebraic* numbers, those that satisfy some polynomial equation with integer coefficients, from *transcendental* numbers, those that don't. It was widely believed that special numbers such as e and π are transcendental, but proofs of these results were some way in the future. In fact, for a time it wasn't even known whether transcendental numbers exist. Joseph Liouville proved that they do, by an explicit construction based on approximations to algebraic numbers. Cantor proved they exist without constructing any.

This way of thinking led Cantor to a broad theory of the number concept, founded in what he called *Mengenlehre*: the theory of collections. We now call it set theory. His work on this theory was mainly published between 1874 and 1884, with further contributions until 1897. It's now seen as the culmination of a century of effort to define 'number' logically and precisely, and as a logical foundation for the whole of mathematics. However, set theory is so basic that its concepts hardly look like mathematics at all, and it certainly didn't in Cantor's day. It's abstract, and it feels *alien* to anyone with the traditional 19th-century mathematical upbringing. Cantor was well aware of this, writing: 'I realise that in this undertaking I place myself in a certain opposition to views widely held concerning the mathematical infinite and to opinions frequently defended on the nature of numbers.'

His ideas were too revolutionary for many; several leading mathematicians denounced them as nonsense, often in robust terms. The influential Leopold Kronecker publicly called Cantor a scientific charlatan, a renegade, and a corrupter of youth. But Kronecker had an axe to grind; he was a number theorist with a rather extremist view of what was permissible in mathematics, famously stating that 'God made the integers, all else is the work of Man.' Today no aspect of mathematics is considered to be God-given, and the logical difficulties that beset the foundations of mathematics already occur for the integers. If the mathematics of the integers is logically consistent, then so is that of the real numbers, complex numbers, and indeed Cantor's theory of sets.

Even in Cantor's day, several leading mathematicians had enough imagination to grasp the magnitude and importance of Cantor's innovation. The most prominent was David Hilbert, who stated that 'No one will expel us from the paradise that Cantor has created.' Eventually Cantor won the argument hands down, but by then he was dead. He suffered from chronic depression from 1904 onwards, and died in a sanatorium in 1918.

Counting and matching

The road to Cantor's paradise begins with the method we use to find out how many things we have: we count them (see Figure 30). A shepherd with a small flock of sheep points at them in succession and chants 'one, two, three, four, five, six, seven'. Running out of sheep, and being careful not to count any of them twice, she concludes that she has seven sheep.

A Maori, faced with the same task and the same flock, would count them too, but the chant would be different: 'ta'i, rua, toru, 'ā, rima, ono, 'itu'. She has 'itu sheep.

Geoffrey Chaucer would have counted 'oon, two, thré, fowre, five, syxe, sevene'. Chaucer had sevene sheep.

There's no disagreement here; just different names for the same things. An astronomer could count the sheep by chanting 'Mercury, Venus, Earth, Mars, Jupiter, Saturn, Uranus'.

Now suppose every sheep is wearing exactly one collar. How many collars are there? No one bothers to count. Instantly the answers ring out: 'Seven!' 'Itu!' 'Sevene!' 'Uranus!' The numbers—where the chant stops—are the same. Why? Because every number matches a sheep, and every sheep matches a collar; so every number matches a collar. The old mathematical name for this procedure is *one-to-one correspondence*. That's a bit of a mouthful, so schools now teach young children about *matching* sets in the hope that this will help them understand numbers. Mathematicians call the procedure a *bijection* or *one-to-one onto mapping*.

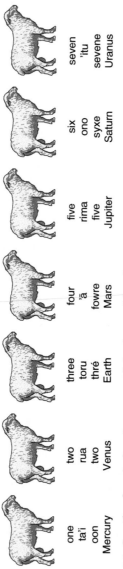

one	two	three	four	five	six	seven
ta'i	rua	toru	'ā	rima	ono	'itu
oon	two	thré	fowre	five	syxe	sevene
Mercury	Venus	Earth	Mars	Jupiter	Saturn	Uranus

30. Counting sheep in three languages and the planets of the solar system.

The principle applies more generally. If the shepherd knew she had 150 sheep, and every sheep was wearing exactly one collar, she could be confident that they had 150 collars. Even if she didn't have a clue how many sheep she had, but knew every sheep was wearing exactly one collar, she could be sure that the number of collars was the same as the number of sheep. This may sound trite, but it's philosophically and foundationally deep. *You can be confident that two numbers are equal without knowing what they are.* There's a sense in which 'having the same number' is more basic than 'how many?'

Around 1880 Gottlob Frege developed a definition of 'number' based on the same principle. He felt that using some standard sequence such as 'one, two, three, four' was arbitrary and inelegant. Any matching sequence would do the same job. Faced with the difficulty of selecting a specific set to represent 'seven', he hit on the cunning idea of using *all of them*. A number, he said, is the set of all sets that match a given set. Then they all match each other, and no set that matches any of them gets left out. It's democratic, inclusive, and unique.

Also, logically flawed. In 1903, just after Frege published the second volume of his masterwork *Grundgesetze der Arithmetik* (basic laws of arithmetic), Bertrand Russell famously demolished a key assumption behind this approach: that phrases of the form 'the set of all sets such that…' are meaningful. His example was 'the set of all sets that do not contain themselves'. If it does, it doesn't; if it doesn't, it does. Contain itself, that is. This is the Russell paradox, a mathematical version of the barber who shaves everyone who doesn't shave themselves, but more precisely stated and immune to clever get-out clauses like female barbers.

These niceties, aside, Frege had a great idea. No need to discard it just because it doesn't provide a useful definition of 'number'. It does provide a useful definition of '*same* number', without any need to say what a number is. Eventually Cantor made it the foundation of this theory of transfinite cardinals.

Cantor's transcendence proof

Cantor didn't develop set theory and his theories of transfinite cardinals off the top of his head. They emerged from his research into standard mathematical questions, and it took him several years to sort out the underlying ideas systematically.

It all started in 1874, when he found a new way to solve a question about transcendental numbers. Recall that algebraic numbers satisfy a polynomial equation with integer coefficients; transcendental numbers don't. For example, $\sqrt{2}$ satisfies the equation $x^2 - 2 = 0$, with integer coefficients 1 and -2, so $\sqrt{2}$ is algebraic. No one had found an algebraic equation for π or e, so they were thought to be transcendental. We now know they are, but at the time their status was conjectural. The first breakthrough came in 1844, when Liouville showed that the error in any rational approximation to an algebraic number must be large, in a technical sense. Therefore a number for which the errors are small must be transcendental. His examples were numbers like

$$0 \cdot 1100001000000000000000001000 \ldots$$

where 1 appears only in positions 1, 2, 6, 24, ... successive factorials. This proved that transcendental numbers exist, but the examples were somewhat artificial.

Cantor solved the problem in a different manner: he proved that there are more transcendental numbers than algebraic ones, in a strong but radically new sense. As a warm-up, I'll prove that the rational numbers are countable: they can be matched with the natural numbers. The proof, which is essentially Cantor's, is similar to the discussion of Hilbert's hotel in Chapter 1. For simplicity, I'll restrict attention to positive rationals, but it's straightforward to extend it to all rationals. The idea is to arrange the positive (which implies non-zero) rationals in order of complexity, where the complexity of p/q is $p + q$. Only finitely many positive rationals have a given complexity, and it doesn't much matter in which order we arrange those, but for definiteness we can start with the smallest p and work up. Finally, we use only fractions in lowest terms—p and q have no common factors greater than 1—to avoid duplication.

The list starts like this:

> *Complexity 2*: 1/1
> *Complexity 3*: 1/2   2/1
> *Complexity 4*: 1/3   3/1
> *Complexity 5*: 1/4   2/3   3/2   4/1
> *Complexity 6*: 1/5   5/1

Then we match them up to whole numbers like this:

1/1	1/2	2/1	1/3	3/1	1/4	2/3	3/2	4/1	1/5	5/1
↕	↕	↕	↕	↕	↕	↕	↕	↕	↕	↕
1	2	3	4	5	6	7	8	9	10	11

and so on. Every rational number corresponds to exactly one natural number, and conversely. With respect to this order (*not* the natural order by numerical value) we can meaningfully refer to the nth (positive) rational number.

To establish the existence of transcendental numbers, Cantor first proves that, like the rationals, the algebraic numbers can be matched, one-to-one, with the natural numbers. His proof is based on a notion of complexity, determined by the largest power of x in the equation defining the algebraic number, and the sizes of the coefficients in the equation. Only finitely many of algebraic numbers occur at a given level of complexity, so we can arrange the algebraic numbers in a list, starting with those that solve an

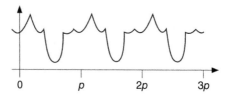

31. Graph of a periodic function with period *p*.

equation of complexity 1, then those that solve one of complexity 2, then complexity 3, and so on. Then we can assign a natural number to any algebraic number by counting along the list.

Finally, Cantor clinches the deal by proving that the real numbers can't be listed in that way. Assume, for a contradiction, that they can. It's now easy to construct a sequence of intervals, each inside the previous one, so that the nth interval does not contain the nth algebraic number. The intersection of those intervals contains some real number. However, this can't be on the list: if it were, it would be the nth algebraic number for some n, but this is excluded from the nth interval, so it can't lie in their intersection. But allegedly every real number is in the list. Therefore no such list exists, and there must be a real number that's not algebraic.

This 'counting' proof was a jaw-dropping moment for Cantor's contemporaries. Either you get it or you don't. If you don't, you think it's nonsense. If you do, you think it's amazingly original and clever. It proves there must be a rabbit in the conjurer's hat—without showing you the rabbit.

Fourier analysis

Cantor's earliest mathematical publications were on number theory. Then he took an interest in a major unsolved problem in Fourier analysis. When Joseph Fourier was developing a mathematical theory of heat flow, he developed a technique for studying periodic functions—functions that repeat the same values over and over again (Figure 31). This distance between successive repetitions is the *period*.

The most familiar periodic functions are the sine and cosine functions, with period 2π. Fourier's idea was to write a periodic function with period 2π as a sum of infinitely many sine and cosine functions:

$$f(x) = a_0 + a_1 \sin x + a_2 \sin 2x + \ldots + b_1 \cos x + b_2 \cos 2x + \ldots$$

He also derived a formula for the coefficients a_0, a_1, a_2, b_1, b_2 in terms of integrals. It wasn't a totally original idea—Euler and Daniel Bernoulli had been arguing about its finer points for some time—but Fourier made effective use of it in his theory of heat, which *was* new. He claimed that *any* 2π-periodic function can be represented by such a series, but Euler and Bernoulli already knew this isn't completely correct. Today we give Fourier credit for his visionary ideas, but criticize his lack of rigour.

The main issue is the convergence of the trigonometric series: whether it has a well-defined sum, and if so, what properties that has. Cantor's research led him to consider the set S of zeros of the series—values of x for which its sum vanishes. The set of zeros can be infinite, and very complicated. It contains a special subset, called the derived set S_1, which consists of all limit points of S. These are the points that are limits of convergent sequences taken from S. Cantor constructed a related Fourier series with zero-set S_1. He noticed that S_1 has its own derived set, S_2, and the process continues with S_3, S_4, and so on.

Crucially, it doesn't stop there. The intersection S_ω of all of these derived sets—the set of points belonging to all of them—need not be empty. If so, it also has a derived set $S_{\omega+1}$, and so on with $S_{\omega+2}$, $S_{\omega+3}$, ... Cantor found an example where this process continued

111

much further. In some sense the structure here is governed by 'infinite numbers' following after all the finite ones:

$$1 \quad 2 \quad 3 \quad \ldots \quad \omega \quad \omega+1 \quad \omega+2 \quad \omega+3$$

Cantor was intrigued by this structure. It's not just fantasy: it's unavoidable if you want to understand the convergence and uniqueness properties of Fourier series. It leads not to transfinite cardinals, but to a related concept, transfinite ordinals. I'll come back to those shortly.

Set theory

Cantor's research on Fourier series naturally led him to basic ideas of set theory and point-set topology, in order to solve a problem that Euler and Bernoulli would have considered mainstream mathematics. It was, but at the time, the tools needed to answer it weren't.

The basic ingredients of set theory are so simple that it doesn't look much like mathematics. A *set* is a collection of objects; in principle any objects, but in practice mathematical ones like numbers or triangles. These objects are its *members* or *elements*. Sets can be combined and manipulated; for example the union of two sets is what you get when you merge them, and the intersection is the set of all members that they have in common.

A finite set can be specified by listing its members, enclosed in braces {}: for example

$$\{1, 2, 3, 4, 5, 6\}$$

is the set of all whole numbers ranging from 1 to 6, and

$$\{2, 3, 5, 7, 11\}$$

is the set of all primes in the range from 2 to 12. Their union is

$$\{1, 2, 3, 4, 5, 6, 7, 11\}$$

and their intersection is

$$\{2, 3, 5\}$$

An infinite set can't be defined like that, but it can be specified by stating what properties its members must have. For example

$$\mathbf{N} = \{n : n \text{ is an integer and } n \geq 0\}$$
$$\mathbf{R}^2 = \{(x,y): x \text{ and } y \text{ are real numbers}\}$$

specify infinite sets. By the way, $\mathbf{R}^2$ is the set-theoretic specification of Euclid's plane, based on coordinates.

Cantor worked out many of the basics of set theory; not as an abstract exercise in formal reasoning, but because he needed it in his work on Fourier series and transcendental numbers. The more he delved into this new topic, the more fascinating and unorthodox his viewpoint became. *Too* unorthodox for many, and we can sympathize with them, because it really does require a new mindset and the rejection of ingrained philosophical principles. To make matters worse, set theory was intimately bound up with a host of other issues that mathematicians had never really sorted out logically. It was a period of great intellectual confusion.

Transfinite cardinals

As well as developing the formalism of set theory, Cantor generalized the notion of cardinal number to arbitrary sets. The set $S = \{1, 2, 4\}$ has 3 members; its cardinal $|S|$ is 3. The set $\mathbf{N}$ of natural numbers has infinitely many members; its cardinal $|\mathbf{N}|$ is…what? Infinity, in some sense—but what sense?

From Frege, and others of his period, Cantor took one key principle. Two finite sets have the same number of members if they can be matched in a one-to-one fashion. The lovely thing about this statement is that you don't need to know what the number of members is. Just as you can check that two sticks have the same length by laying them side by side, without measuring the actual lengths.

Cantor realized you can do the same thing for infinite sets. He introduced a new kind of infinite number, often said to be *transfinite*, and defined it to be the cardinal of **N**. He gave it a symbol: not ∞, which was potentially confusing given the huge variety of distinct usages of that symbol in mathematics, but $\aleph_0$. Here $\aleph$ is 'aleph', the first letter in the Hebrew alphabet. Following Frege's lead, any set that can be put in one-to-one correspondence with **N** is also assigned the cardinal $\aleph_0$. The integers (positive and negative) are an example. One way to define the correspondence is to interleave positive and negative integers like this:

0	1	2	3	4	5	6	7	8	9	10	11	12
$\updownarrow$	$\updownarrow$	$\updownarrow$	$\updownarrow$	$\updownarrow$	$\updownarrow$	$\updownarrow$	$\updownarrow$	$\updownarrow$	$\updownarrow$	$\updownarrow$	$\updownarrow$	$\updownarrow$
0	1	−1	2	−2	3	−3	4	−4	5	−5	6	−6

Another example is the positive rationals: list them in order of complexity, as already described. To include negative rationals as well, interleave them with the positive ones in a similar manner to the negative integers above. Yet another is Galileo's remark about perfect squares (see Chapter 1). The set of all squares can be matched to **N**, so it also has cardinal $\aleph_0$.

Are there any sets that are neither finite nor have cardinal $\aleph_0$? From his work on transcendental numbers, Cantor knew there is: the set **R** of real numbers. He expected this to be the next cardinal bigger than $\aleph_0$, in which case it could be defined to be a new cardinal $\aleph_1$, no doubt followed by $\aleph_2$, $\aleph_3$, $\aleph_4$, and so on—an endless series of transfinite numbers. Indeed, he proved that

32. A one-to-one mapping from the open unit interval onto R.

there's no largest cardinal. The set of all subsets of any given set must have a larger cardinal than the set itself. That is, there's no way to match every subset to a member of the set in a one-to-one manner. The proof, ironically, is a variation on the idea behind the Russell paradox.

One reason why many found Cantor's ideas counterintuitive is that when you match two sets, most of their traditional features are irrelevant. The natural order of numbers can be jumbled up, for example. This is why we can match rationals to natural numbers. The dimensions of spaces are an unnecessary encumbrance, which is why the real line **R** and the plane $\mathbf{R}^2$ match—a result that astonished even Cantor when he first proved it, because the plane looks so much larger than the line. Order, dimension, algebraic operations, and the like are extra mathematical superstructure attached to bare sets. The superstructure can itself be defined using set theory, but it's not automatically built into the underlying set.

Cantor used operations on sets to define arithmetical operations on transfinite cardinals—sum, product, exponential. He established their basic properties. He and other mathematicians also defined 'greater than' and 'less than'. He already knew from his work on transcendental numbers that the cardinal of **R**—call it **c**—is greater than $\aleph_0$, but he couldn't prove that **c** is the smallest cardinal with that property. Is there a cardinal strictly between $\aleph_0$ and **c**? If not, it makes sense to define $\aleph_1$ to be **c**. Plenty of sets contain **N** but are contained in **R**—the integers, rationals, algebraic numbers, positive real numbers, transcendental numbers, and an endless host of others. These are sensible candidates for an intermediate cardinal, but in every case the cardinal is either $\aleph_0$ or c. The integers, rationals, and algebraic numbers all have cardinal $\aleph_0$; the positive reals and the transcendental numbers have cardinal c.

Obvious candidates turned out not to work. For example, surely there are fewer real numbers between 0 and 1 than there are reals in total? It seems plausible: the length of the interval from 0 to 1 (excluding the ends, for definiteness) is 1; the length of $\mathbf{R}$ is infinite. However, one-to-one correspondences are no respecters of length. The graph $y = (1 - 2x)/(x^2 - x)$, shown in Figure 32, maps the unit interval $(0,1)$ in one-to-one manner onto the whole of $\mathbf{R}$.

The statement that $\aleph_1 = c$ became known as the Continuum Hypothesis. Hilbert listed it among his famous twenty-three unsolved problems in 1900. Some mathematicians began to suspect that the statement was related to foundational issues in mathematical logic. Gödel proved in 1940 that the truth of the Continuum Hypothesis is logically consistent with the standard axiomatic formulation of set theory, known as the Zermelo–Fraenkel axioms. Finally, in 1963, Paul Cohen proved that the falsity of the Continuum Hypothesis is *also* logically consistent with the Zermelo–Fraenkel axioms. It's a stunning example of a statement that is independent of the usual axioms. There are versions of set theory in which the Continuum Hypothesis is true, but there are also versions of set theory in which the Continuum Hypothesis is false.

Precursors to Cantor

Cantor's work is profound because he set up a logically rigorous framework, defined infinite analogues of counting and numbers, and proved that these concepts have specified properties. However, he wasn't the first person to suggest that infinity can come in different sizes. That honour, as far as historians of mathematics are aware, goes to an unknown Indian mathematician or mathematicians around 400 BC. The suggestion is documented in *Surya Prajnapti*, a Jain mathematical text.

We saw in Chapter 2 that like many Indian religions, Jainism was fascinated by very large numbers. Moreover, they were aware that no counting number, however large, is infinite. That honour they reserved for the smallest uninnumerable number. Beyond this, they asserted, stretch even larger infinite numbers. They classified numbers into three types, each with three subtypes:

Enumerable: lowest, intermediate, and highest.
Innumerable: nearly innumerable, truly innumerable, and innumerably innumerable.
Infinite: nearly infinite, truly infinite, and infinitely infinite.

They also distinguished five different meanings of 'infinity': infinite in one direction, infinite in two directions, infinite in area, infinite everywhere (that is, in volume), and perpetually infinite.

Cantor had a similar vision, but he fleshed it out with rigorous definitions. His conclusions were similar, but with important differences. For example, in his formulation a line, a plane, and a volume all have the same number of points. This doesn't mean the Jains were wrong. They were thinking about subtly different ideas. They were two millennia ahead of their time, but their views on the infinite were somewhat mystical, and not formulated with the logical precision we now require.

After the Jains, the next significant mathematical contribution to our understanding of the infinite was probably Galileo's. Chapter 1 includes an extract from *Two New Sciences*, in which Salviati, the irritating know-all, argues that 'there are as many squares as there are numbers', because 'every number is the root of some square'. Simplicio, the dunce, agrees; he wouldn't stand a chance if he didn't. Sagredo, the straight man, feeds Salviati useful lines when the performance starts to flag.

The philosophical issue that Galileo was addressing here is the belief that 'the whole is greater than the part'. There are obviously more numbers than squares, because every square is a number but some numbers are not square. Let's take a look, marking squares in boldface:

$$\mathbf{1}\ 2\ 3\ \mathbf{4}\ 5\ 6\ 7\ 8\ \mathbf{9}\ 10\ 11\ 12\ 13\ 14\ 15\ \mathbf{16}\ 17$$

The proportion of squares gets steadily smaller, except when we meet the next one. Among the numbers up to 100 there are 10 squares; up to 10,000 there are 100 squares; up to a million there are only a thousand squares.

Salviati's point is that although squares get thinner and thinner on the ground, they never run out altogether. Sooner or later, another one comes along. So we can match the numbers to the squares:

1	2	3	4	5	6	7	8	9	10	11	12
$\updownarrow$	$\updownarrow$	$\updownarrow$	$\updownarrow$	$\updownarrow$	$\updownarrow$	$\updownarrow$	$\updownarrow$	$\updownarrow$	$\updownarrow$	$\updownarrow$	$\updownarrow$
1	4	9	16	25	36	49	64	81	100	121	144

If we stop at some finite limit n, there are a lot more numbers than squares. But if we don't stop, every number matches exactly one square, and conversely. So the part can match the whole. Cantor's ideas explained Salviati's observation.

Transfinite ordinals

There are two ways to view the whole numbers: as cardinals, numerical measures of how big something is, or as ordinals, which place objects in order by running through the sequence 1, 2, 3, ... For finite numbers this distinction is a bit nit-picking, and makes

118

precious little difference. When we come to infinite numbers, however, there's a big difference. To each transfinite cardinal there correspond infinitely many different transfinite ordinals. Cantor first ran into transfinite ordinals in his work on Fourier series.

Transfinite cardinals satisfy the well-ordering principle: any set of cardinals has a smallest member, necessarily unique. There's a parallel theory of well-ordered sets, in which the one-to-one correspondence is required to preserve order as well. Now cardinals are replaced by ordinals, and the smallest infinite ordinal, corresponding to **N**, is called ω. Any well-ordered set that can be placed in one-to-one correspondence with **N**, *without disturbing the order*, has ordinal ω. An example is Galileo's set of squares: the correspondence keeps them in their natural order. Another is the set of all primes.

Ordinals have a rich structure with some strange features.

In cardinal arithmetic, $\aleph_0 + 1$ is equal to $1 + \aleph_0$, and both are equal to $\aleph_0$. Suppose we take the natural numbers **N** and add a new element X, different from any natural number. Because the ordering is not important for cardinals, we can match this larger set to **N** by shifting everything in **N** along one space, and putting X at the beginning, just like accommodating a new guest in Hilbert's hotel:

$$
\begin{array}{ccccccccccccc}
0 & 1 & 2 & 3 & 4 & 5 & 6 & 7 & 8 & 9 & 10 & 11 & 12 \\
\updownarrow & \updownarrow & \updownarrow & \updownarrow & \updownarrow & \updownarrow & \updownarrow & \updownarrow & \updownarrow & \updownarrow & \updownarrow & \updownarrow & \updownarrow \\
X & 0 & 1 & 2 & 3 & 4 & 5 & 6 & 7 & 8 & 9 & 10 & 11
\end{array}
$$

This reasoning applies to both $\aleph_0 + 1$ and $1 + \aleph_0$, again because order is unimportant. Transfinite cardinals satisfy the commutative law of addition $a + b = b + a$. By similar reasoning, we can prove theorems such as:

$$\aleph_0 + 2 = \aleph_0 \qquad \aleph_0 + \aleph_0 = \aleph_0 \qquad 1066\aleph_0 = \aleph_0$$
$$\aleph_0^2 = \aleph_0 \qquad \aleph_0^3 = \aleph_0 \qquad \aleph_0^{1066} = \aleph_0$$

and so on. However, some arithmetical operations lead to larger cardinals; for instance,

$$2^{\aleph_0} > \aleph_0$$

Indeed, $2^{\aleph_0} = c.$

When it comes to ordinals, the rules are very different. For instance, $\omega + 1$ is not equal to ω, but larger. A set with ordinal $\omega + 1$ can be constructed by taking the natural numbers **N** and appending a new element X, deemed to be greater than any natural number; that is, coming after them all in the ordering. So this ordered set looks like

$$1 \quad 2 \quad 3 \quad 4 \quad 5 \quad 6 \quad 7 \ldots X$$

with X tagged on as an afterthought. This is less artificial than it might appear, since the set of all cardinals up to and including $\aleph_0$, in order of size, looks like

$$1 \quad 2 \quad 3 \quad 4 \quad 5 \quad 6 \quad 7 \ldots \aleph_0$$

Indeed, we can write the sequence of all possible distinct cardinals as $\aleph_\alpha$, where α runs through all ordinals in ascending order.

When dealing with ordinals, we can no longer do a Hilbert hotel and match this larger set to **N** by shifting everything in **N** along one space and putting X at the beginning, because now we have to keep everything in the same order. Since X is the largest element, we can't move it to the front, where it would become the smallest. We can't put it anywhere in the middle, somewhere inside **N**, for the same reason. It has to remain where it is. So $\omega + 1$ is different from ω. In fact, it's the next biggest ordinal after ω.

The sequence of infinite ordinals goes like this:

$$\omega \quad \omega+1 \quad \omega+2 \quad \omega+3 \quad \omega.2 \quad \omega.2+1 \ldots \omega^2 \ldots \omega^3 \ldots \omega^\omega \ldots$$

and so on. If we ignore the order, the corresponding sets can all be made to match **N**, so they *all* have cardinal $\aleph_0$. But eventually we reach the first uncountable ordinal, denoted by ε_0.

On the other hand, $1+\omega$ is the same as ω, because now we can add an extra element X at the front and shift **N** along one space without disturbing the relative order, which is how the first extra guest got a room in Hilbert's hotel. So addition of ordinals doesn't satisfy the commutative law, unlike cardinals.

We can now return to the first example in Chapter 1: the meaning of $\infty+1$. It depends on how we interpret the symbol ∞. If it's the transfinite cardinal $\aleph_0$, then $\aleph_0+1=\aleph_0=1+\aleph_0$. If it's the transfinite ordinal ω, then $\omega+1>\omega$ but $1+\omega=\omega$. If it's $1/\varepsilon$, where ε is Cauchy's infinitesimal sequence $(1/n)$, then $1/\varepsilon+1=1+1/\varepsilon>1/\varepsilon$. Each meaning must be investigated in its own right.

Cantor and Wittgenstein

Cantor repeatedly emphasized that set theory was about actual infinity. He explicitly contrasted it with Aristotelian potential infinity, and he discussed various philosophical views about infinity in his work. But the issues are rather subtle.

If you interpret Cantor's ideas literally, they refer to a (conceptual) actual infinity. We think of the set **N** of all natural numbers as an object, not as a process. In Cantor's view, a set, be it finite or infinite, is a valid mathematical object. The set of all natural numbers 'exists' in the same way that {1, 2, 3} exists. Cantor aimed to associate with each infinite set a cardinal, determining how many members it has. It's difficult to see how to state these ideas using only the language of potential infinity. At best, any attempt would be hopelessly contrived.

This is especially clear for a second proof that the real numbers are uncountable, which Cantor gave in 1891. It's more elementary and avoids assertions about nested sequences of closed intervals. It starts by assuming the set **R** of real numbers is countable, and derives a contradiction. This is obtained by first reducing the issue to a similar statement about real numbers greater than 0 and less than 1, which is routine using Figure 32. Having done this, every number in that range has a decimal expansion

$$x = 0 \cdot x_1 x_2 x_3 x_4 \ldots$$

Such an expression isn't quite unique; for example $0 \cdot 199999 \ldots$ is equal to $0 \cdot 2$. (People often think these are different, by an infinitesimal amount, but in conventional mathematics they're the same. Just as 1/2 and 2/4 look different, but represent the same fraction.) To remove the ambiguity, forbid infinite recurring sequences of 9s.

Suppose **R** is countable. Then the counting numbers **N** can be matched to **R**:

1	$0\ a_1 a_2 a_3 a_4 \ldots$
2	$0\ b_1 b_2 b_3 b_4 \ldots$
3	$0\ c_1 c_2 c_3 c_4 \ldots$
4	$0\ d_1 d_2 d_3 d_4 \ldots$

By assumption, every real number occurs somewhere in the list. Now we construct one that doesn't. We define successive decimal places $x_1, x_2, x_3, \ldots$ of this number x as follows:

If $a_1 = 0$ let $x_1 = 1$. If $a_1 > 0$ let $x_1 = 0$
If $b_2 = 0$ let $x_2 = 1$. If $b_2 > 0$ let $x_2 = 0$
If $c_3 = 0$ let $x_3 = 1$. If $c_3 > 0$ let $x_3 = 0$
If $d_4 = 0$ let $x_4 = 1$. If $d_4 > 0$ let $x_4 = 0$

and so on. In general, make x_n either 0 or 1, and different from the nth digit of the real number corresponding to n.

By construction, x differs from every number on the list. It differs from the first number in its first digit, from the second number in its second digit; in general, it differs from the nth number in its nth digit, so it's different from the nth number, no matter what value n has. However, we assumed that the list exists, and every real number appears on it. This is a contradiction, and what it contradicts is the assumption that such a list exists. Therefore no such list exists, and **R** is uncountable.

Wittgenstein despised the diagonal argument. In *Lectures and Conversations* he offered to put the proof in such a way that 'it will lose its charm for a great number of people and certainly will lose its charm for me'. In *Remarks on the Foundations of Mathematics* he disputed Hilbert's 'paradise' remark. As for no one being expelled, his *Lectures on the Foundations of Mathematics* claimed that 'you'll leave of your own accord'. With Cantor long dead, Wittgenstein continued to express his profound philosophical dissatisfaction, complaining that mathematics was 'ridden through and through with the pernicious idioms of set theory'.

Setting 'pernicious' aside, it was, and still is. No one left Cantor's paradise. A few decided not to enter, but those that did found little to justify Wittgenstein's scepticism. Cantor's new-found freedom has taken mathematics from strength to strength. Hilbert was right.

Mathematics, philosophy, and religion

Cantor's approach does raise a major philosophical issue. It requires us to think of the sets involved as specific objects, not processes: actual infinity, in Aristotle's sense. 'Actual' in a conceptual manner, of course, as for all mathematical concepts, a distinction that wasn't fully understood in Aristotle's time. Cantor had a deep interest in philosophy, and was well aware of the explosive nature of this view, but he found it impossible to avoid.

At that time, less than 150 years ago, the concept of actual infinity was common ground (often battleground) for mathematics, philosophy, and religion. Europe was intensely Christian and belief in God was the default view, although atheism and agnosticism were already beginning to gain ground. Christians saw their deity as a perfect, infinite, eternal being; indeed, as the *unique* actually infinite being. They had no qualms about Aristotle's potential infinity, but asserting the existence of another *actual* infinity was theological dynamite—even when the 'actual' infinity concerned was a mathematical abstraction.

Religion's political grasp was slipping, so the churches didn't make as much fuss as they had in the 17th century about infinitesimals, but the status of mathematical infinity was a serious issue for the religious. When Kronecker said 'God created the integers' he wasn't speaking metaphorically. Cantor was also religious, and he went to considerable lengths to explain how, in his view, transfinite cardinals could be reconciled with God as the unique absolute infinity. He could prove there is no largest transfinite number. Paradoxically, this implies that the set of all transfinite numbers (surely a meaningful and indeed important set in his theory) is so big that it doesn't have a cardinal. Impressively infinite though any given transfinite cardinal may be, it can't approach the absolute infinity of God. This justification has clear echoes of Augustine's proof that God is infinite (see Chapter 3).

Cantor stated that 'the transfinite species are just as much at the disposal of the intentions of the Creator and His absolute boundless will as are the finite numbers', neatly turning the tables on some critics. If you claim that transfinite cardinals are on a par with God, then you're saying there are limits to His power, which is theologically suspect. Cantor believed that his knowledge of transfinite cardinals had come directly from God, and it was his Christian duty to tell the world about them. He corresponded with distinguished philosophers and theologians, and published the correspondence. He even wrote to Pope Leo XIII, and sent him several pamphlets on the topic.

This may seem extreme, but such was the spirit of the age. It didn't help that foundational issues in mathematics had stimulated several schools of thought that rejected the infinite, or accepted it only when a specific *construction* could be given. Kronecker was a constructivist of this kind. Liouville's work on transcendental numbers, which constructed a specific example, was acceptable; Cantor's alleged proof that transcendentals exist didn't actually construct one, so it was rubbish. In fact, even the logical basis of mathematics was under attack by the Intuitionist school of Luitzen Brouwer, which rejected proof by contradiction as well.

A version of constructivism still exists. Founded by Errett Bishop in his 1967 *Foundations of Constructive Analysis*, it seeks constructive analogues of basic mathematical theorems. For example, the intermediate value theorem of standard mathematics states that if a continuous function is negative at some point and positive at another, then somewhere in between it must be zero. Constructive analysis rejects the usual existence proof, insisting it must be replaced by an algorithmic procedure that defines such a point explicitly. However, there's a price to pay: the constructive analogue of 'continuous' has to be a much stronger property than the traditional one, so the analogy is incomplete. Most mathematicians view constructive analysis as a valid but specialized area of mathematics, focused on constructions and algorithms, proving analogues of basic theorems in which stronger hypothesis give stronger results. Constructivists, however, tend to see it as a replacement for existing mathematics; its theorems are not analogues, but the only logically valid way to proceed. This view has not made much headway.

The idea that we can't be sure something exists unless we're told exactly how to find it has a definite appeal. If David Livingstone had come back from Africa saying 'I've proved that the source of the

Nile exists', no one would have been impressed. They wanted to know where the source was, and that's what Livingstone told them. (He was wrong, but no matter.) However, existence arguments cut through a lot of irrelevant detail, and we use them more than we imagine. If a friend is walking along the road to your house, and you head off along the reverse route to intercept him, you're confident you'll meet up, even though you can't predict precisely where.

Vicious narrow-minded opposition caused Cantor much grief. In 1904 Julius König gave a lecture at the 3rd International Congress of Mathematicians, claiming that transfinite numbers and set theory were based on an error. König was wrong; Ernst Zermelo found a mistake in his alleged proof the very next day. But Cantor had been humiliated in public, and was so distressed that he even began to doubt his faith. This event may have triggered his recurrent depression. He retired just before World War I, lived in poverty, and died ten months before the war ended.

Ironically, Cantor's own view of infinitesimals, as formalized by du Bois-Reymond and Stolz (Chapter 4), was just as extreme. He called infinitesimals the 'cholera bacillus of mathematics'. He probably objected to the approach to infinity, which differs significantly from his own. The reciprocal of an infinitesimal doesn't sit comfortably with counting. The verdict of posterity is that Cantor was wrong; he ignored his own views on mathematical freedom and he forgot that 'infinity' in mathematics can have many meanings, not just the one he was advocating.

Processes and things

Set theory habitually reifies processes as objects. For example a mathematical function was traditionally considered to be a process: a rule for transforming an 'input' into a related 'output', both usually being numbers. The 'square' function turns any input into its square. This can be considered an instance of potential infinity, because the rule itself can be stated in finite terms:

'multiply the number by itself', and it can then be applied to any finite collection of numbers without invoking anything infinite.

However, the standard definition of this function in set theory is: the set of all pairs of the form (x, x^2). This is a conceptual table of all values of the function, just like normal logarithmic or trigonometric tables, except that all possible values of x are listed and both x and x^2 are arbitrary real numbers, determined to infinite precision. The list is thought of as a completed 'actual' object, not as the process defining that object. The process is replaced by 'look it up in the table'. From a foundational viewpoint, all functions are now defined in this manner.

As Wittgenstein saw, many uses of this construction can be reduced to properties of the process itself, so they can be rephrased in terms of potential infinity. For example, in Cantor's theory, the set of all whole numbers and the subset of all even numbers can be put into one-to-one correspondence by taking the input n to the output $2n$. This construction (which we used to accommodate a coachload of tourists in Hilbert's jam-packed hotel in Chapter 1) was then viewed as a proof that the whole can equal the part. In *Philosophical Remarks*, Wittgenstein correctly realized that actual infinity isn't essential to this particular construction, saying: 'Does the relation $m = 2n$ correlate the class of all numbers with one of its subclasses? No. It correlates any arbitrary number with another, and in that way we arrive at infinitely many pairs of classes, of which one is correlated with the other.'

Fair enough: he's pointing out that all you need is the rule for getting output m from input n. But few mathematicians would concur with what follows: 'but which are never related as class and subclass. Neither is this infinite process itself in some sense or other such a pair of classes... In the superstition that $m = 2n$ correlates a class with its subclass, we merely have yet another case of ambiguous grammar.'

Wittgenstein hated set theory with a passion, declaring that it had 'completely deformed' philosophical and mathematical attitudes. Whatever the merits of his views, he was disappointed. Mathematicians ignored his advice, with good reason. Even if he was right, and any apparent use of actual infinity can be recast in finite form, that's not an argument for setting aside the infinite. On the contrary, it tells us that the infinite introduces no logical inconsistencies that aren't already present in the finite. The two approaches are not locked into a conflict that tears them apart: they constitute an alliance that strengthens both. Moreover, translating proofs from the language of the infinite to the language of the finite often turns simple, transparent statements into tortuous, laborious ones. Set theory is a thinking tool, and a very powerful one, because it makes complex ideas simple.

Think about turning the diagonal proof into a potential infinity proof using processes. Each real number is an infinite decimal—a process. The hypothetical list is a process operating on those processes. The missing real number obtained by looking along the diagonal is a process operating on a hypothetical process applied to processes. The contradiction comes by comparing this with each process involved in the process that represents the list. The result might just be philosophically purer, but it would also be utterly incomprehensible.

For reasons like this, set theory has now completely taken over advanced mathematics, both pure and applied, including applications to the physical sciences, biology, even economics. Not just as a way of formulating concepts, or as a notation, but in fully-fledged Cantorian splendour. Different sizes of infinity, especially the distinction between countable and uncountable infinity, are vital to huge areas, and indispensable in applied science. Probability theory rests on the concept of countable additivity of a measure. That is, if you add up a generalization of area for a countably infinite collection of disjoint sets, you get the area of their union. Uncountable additivity, on the other hand, can be proved self-contradictory. Partial differential equations received a huge boost from Stefan Banach's introduction of infinite-dimensional spaces of operators. Quantum mechanics depends on Hilbert spaces, key examples of Banach's new concept.

Wittgenstein made the same category error as Locke. The 'actual infinities' of set theory are mathematical concepts, not real objects. They're 'mathematically actual' when the mathematician contemplates them as completed things, not as processes. They're 'mathematically potential' when the mathematician contemplates them as processes. What distinguishes the two viewpoints, in any specific piece of mathematics, is how they're used. Not what they 'really' are. They aren't anything real.

Mathematical existence

For most mathematicians, making sense of infinity is not about the meaning of the infinite; it's about the meaning of mathematical existence. There's a strong consensus that mathematics isn't reality; it just resembles reality in useful ways. A mathematical object or process *exists* if it doesn't lead to logical contradictions, which is the viewpoint that Cantor promoted. Its existence, in that sense, can be proved by constructing it within the normal framework of known mathematics, or by showing that its non-existence leads to a logical contradiction.

This view of mathematical existence is problematic in one respect: it assumes mathematics itself is logically consistent. If not, the criterion implies that mathematics fails to exist. However, Gödel proved that the consistency of mathematics can never be proved within any axiomatic framework . . . unless it's *false*, in which case anything can be proved. Gerhard Gentzen provided a consistency proof in 1936 based on transfinite ordinals, but that method is of course open to philosophical doubts.

The second type of existence proof—non-existence leads to a contradiction—is non-constructive. Some philosophically minded mathematicians object to such proofs. However, even in everyday life we make common use of non-constructive arguments—usually without noticing. I wonder how constructivists would react if the police gave them a speeding ticket because their average speed, measured over a particular stretch of road, exceeded the speed limit. If it came to court, the police would argue that such an average proves the existence of some time at which the accused's speed exceeded the limit. A sufficiently committed constructivist would be obliged to argue that unless the police can establish a specific time at which they exceeded the limit, there's no case to answer.

I don't want to leave you with the impression that mathematics has explained every puzzle about infinity. Not even that mathematicians *claim* to have explained every puzzle about infinity. There are still plenty of unsolved problems, especially in axiomatic set theory. But mathematicians have put together a logical framework in which we can understand those questions, answer many, and make distinctions between different instances of infinity. That framework has led to dramatic new discoveries, enriching mathematics and leading to new applications.

Welcome to the bizarre but beautiful world of the infinite.

References

Extract from Galileo's 1638 *Discorsi e Dimostrazioni Matematiche Intorno a Due Nuove Scienze*: Galileo Galilei, *Dialogues Concerning Two New Sciences*, translated by Henry Crew and Alfonso de Salvio, Macmillan, New York 1914.

Quotation from Archimedes' *Psammites*: James R. Newman, *The World of Mathematics*, Simon and Schuster, New York 1956.

Quotations by Aristotle about Zeno's paradoxes: Aristotle, *Physics*, translated by R. P. Hardie and R. K. Gaye from *The Complete Works Of Aristotle* (editor Jonathan Barnes), Princeton University Press, Princeton 1984.

Quotation by Paul du Bois-Reymond about the infinitely small: Paul du Bois-Reymond, Über die Paradoxen des Infinitär-Calcüls, *Mathematische Annalen* 11 (1877) 150–67.

Extract from Immanuel Kant, *Critique of Pure Reason*, translated by Paul Guyer and Allen Wood. Cambridge: Cambridge University Press, 1998.

Further reading

Chapter 1: Puzzles, proofs, and paradoxes

Brian Clegg. *Brief History of Infinity: The Quest to Think the Unthinkable*, Robinson, London 2003.

Reviel Netz and William Noel. *The Archimedes Codex*, Weidenfeld & Nicolson, London 2007.

Rudy Rucker. *Infinity and the Mind: The Science and Philosophy of the Infinite*, Princeton University Press, Princeton 2004.

Chapter 2: Encounters with the infinite

Eugene P. Northrop. *Riddles in Mathematics: A Book of Paradoxes*, Penguin, Harmondsworth 1960.

Ian Stewart and David Tall. *The Foundations of Mathematics* (2nd ed.), Oxford University Press, Oxford 2015.

David Foster Wallace. *Everything and More: A Compact History of Infinity*, W.W. Norton, New York 2004.

Chapter 3: Historical views of infinity

John Bowin. Aristotelian infinity, *Oxford Studies in Ancient Philosophy* 32 (2007) 233–50.

Kevin Davey. Aristotle, Zeno, and the stadium paradox, *History of Philosophy Quarterly* 24 (2007) 127–46.

Michael Heller and W. Hugh Woodin (eds.). *Infinity: New Research Frontiers*, Cambridge University Press, Cambridge 2011.

Joe Mazur. *Zeno's Paradox: Unraveling the Ancient Mystery Behind the Science of Space and Time*, Plume, New York 2008.

Chapter 4: The flipside of infinity

Amir Alexander. *Infinitesimal: How a Dangerous Mathematical Theory Shaped the Modern World*, Scientific American/Farrar, Straus and Giroux 2014.

Mikhail Katz and David Sherry. Leibniz's infinitesimals: their fictionality, their modern implementations, and their foes from Berkeley to Russell and beyond, *Erkenntnis* 73 (2013) 571–625.

H. Jerome Keisler. *Elementary Calculus: An Infinitesimal Approach*, University of Wisconsin, Madison 2000.

Abraham Robinson. *Non-Standard Analysis* (2nd ed.), Princeton University Press, Princeton 1996.

Chapter 5: Geometric infinity

Kirsti Andersen. *The Geometry of an Art: The History of the Mathematical Theory of Perspective from Alberti to Monge*, Springer, New York 2007.

H. S. M. Coxeter. *Introduction to Geometry*. John Wiley & Sons, New York 1969.

Morris Kline (ed.). *Mathematics in the Modern World*, W.H. Freeman, San Francisco 1968.

Chapter 6: Physical infinity

Neil J. Cornish, David N. Spergel, Glenn D. Starkman, and Eiichiro Komatsu. Constraining the topology of the universe, *Physics Review Letters* 92 (2004) 201302.

Tim Poston and Ian Stewart. *Catastrophe Theory and Its Applications*, Dover Publications, New York 1996. (Reprint of 1978 Pitman edition.)

Donald Saari and Zhihing Xia. Off to infinity in finite time, *Notices of the American Mathematical Society* 42 (1995) 538–46.

Pascal M. Vaudrevange, Glenn D. Starkman, Neil J. Cornish, and David N. Spergel. Constraints on the topology of the universe: extension to general geometries. *Physical Review D* 86 (2012) 083526.

Chapter 7: Counting infinity

Joseph Warren Dauben. *Georg Cantor: His Mathematics and Philosophy of the Infinite*, Princeton University Press, Princeton 1979.

L. C. Jain. Set theory in the Jaina school of mathematics, *Indian Journal of History of Science* 8 (1973) 1–27.

George G. Joseph. *The Crest of the Peacock: Non-European Roots of Mathematics* (2nd ed.), Penguin, London 2000.

Navjyoti Singh. *Jain Theory of Actual Infinities and Transfinite Infinities*, National Institute of Science, Technology and Development Studies (NISTAD), New Delhi 1987.

Ian Stewart and David Tall. *The Foundations of Mathematics* (2nd ed.), Oxford University Press, Oxford 2015.

Publisher's acknowledgements

We are grateful for permission to include the following copyright material in this book.

Extract from Galilei, Galileo (1954) [1638]. *Dialogues concerning two new sciences*. Translated by Henry Crew and Alfonso de Salvio (New York: Dover 2003). pp. 31–3, with permission.

The publisher and author have made every effort to trace and contact all copyright holders before publication. If notified, the publisher will be pleased to rectify any errors or omissions at the earliest opportunity.

Index

Index

SOCIAL MEDIA
Very Short Introduction

Join our community

www.oup.com/vsi

- Join us online at the official Very Short Introductions **Facebook** page.
- Access the thoughts and musings of our authors with our online **blog**.
- Sign up for our monthly **e-newsletter** to receive information on all new titles publishing that month.
- Browse the full range of Very Short Introductions online.
- Read **extracts** from the Introductions for free.
- If you are a teacher or lecturer you can order inspection copies quickly and simply via our website.